T0227419

# THE

# CallCenter

# DICTIONARY

## The Complete Guide to Call Center & Customer Support Technology Solutions

COVERING

CRM

Telemarketing

Customer Service

Voice Processing

Switches

Software

Training

Call Center Management

CRC Press
Taylor & Francis Group
Boca Raton London New York

CRC Press is an imprint of the
Taylor & Francis Group, an informa business

**Madeline Bodin** and **Keith Dawson**

CRC Press
Taylor & Francis Group
6000 Broken Sound Parkway NW, Suite 300
Boca Raton, FL 33487-2742

First issued in hardback 2017

ISBN 13: 978-1-138-41233-0 (hbk)
ISBN 13: 978-1-57820-095-5 (pbk)

**Visit the Taylor & Francis Web site at**
**http://www.taylorandfrancis.com**

**and the CRC Press Web site at**
**http://www.crcpress.com**

# Introduction

Put this dictionary to work. Keep it by your desk. Throw it in your briefcase as you head out to a meeting. Refer to it when you are trying to figure out what, exactly, that great new technology that someone is trying to sell you is supposed to do.

This dictionary is the updated, expanded and revised edition of a book that was first written in 1996, but it has almost twenty years of experience behind it. We are not merely call center experts, but editors and writers who take our craft seriously. We've spent countless hours thinking about how to make a very difficult subject understandable to the over-worked call center manager trying to keep up with the industry while gobbling down lunch.

As editors and prolific authors on call center issues, we've not only been thinking and learning about call centers for years, but we've put serious effort into the way we explain what we've learned and thought. Our aim has always been to make this information clear, accessible, and, occasionally, fun.

The dictionary you hold in your hands is the fruit of all those years of thought and care.

It's a dictionary you can use to explain things to your customers. A dictionary you can consult after a meeting with the MIS department. Use it to get a leg up on the jargon of your new job or to explore what a call center can do to improve your business.

The language of the call center includes not only terms specific to the industry, but it also relies on terms from telecommunications, computers, networking, training, sales and general business management. But certain terms in these fields are more pertinent than others. Our goal was not to throw in every definition, including the kitchen sink, but to make our dictionary as complete AND concise as possible.

The call center industry continues to evolve. We've added dozens and dozens of new terms in this second edition. Some describe concepts that were unheard of just a few years ago. We've also revised dozens more terms to reflect changes in the industry. We are confident that this dictionary can help you, not only today, but in the months and years to come. If you are puzzled by a term that you don't find here, please let us know. If the lingo is different in your call center, we'd like to know that too. We are eager to hear your suggestions for our 3rd edition of The Call Center Dictionary.

*Madeline Bodin*

Madeline Bodin

*Keith Dawson*

Keith Dawson

CMP Books

# numbers

**0345 NUMBERS** A British Telecom LinkLine service in England where the caller is charged at the local rate irrespective of the distance of the call. The subscriber pays installation and rental charges in addition to a charge for each call.

**0800 NUMBERS** The British equivalent of an 800 number. An 0800 number is a British Telecom LinkLine service in England where the caller is not charged for the call.

**0891 AND 0898 NUMBERS** Sort of like a 900 number. A British Telecom Premium rate service in England where the caller is charged at a premium rate for the call. The calls are normally made to receive information or a service. The service provides revenue for the information provider who receives part of the call charge.

**1+** Pronounced "One plus." In North America, dialing 1 as the first digit signals your local phone company that the phone number you are dialing is long distance, but its destination is one within the North American Numbering Plan. The number 1 will typically be followed by an area code and then seven digits. To reach other international countries, from the United States, you dial the international access code "011."

**1A2** A very basic key telephone system. Often used behind Centrex or a PBX. See KEY SYSTEM.

**1.544 MBPS** The speed of a North American T-1 circuit. See T-1.

**2600 TONE** On a telecommunications network with in-band signaling, the method for signaling that a line is NOT being used is a 2600 Hertz tone. (You can't have nothing on the line, because that "nothing" might be a pause in the conversation.) This tone makes the network ripe for abuse. Just put a 2600 Hz tone on the line, and you get a free call. That's why these days just about the entire telecommunications network uses Signaling System 7 — and out-of-band signaling. Out-of-band signaling is much harder to abuse.

**30-DAY SYNDROME** How the headset industry refers to the sudden change in headset preferences in call center agents and other headset users after a month or so of use. At first, agents are concerned mostly with looks. That is, they don't want to look like dorks while wearing their headsets. They also don't want to get their hair messed up by the headset's headband. After using a headset for hours at a stretch for 30 days, though, all they want is comfort and clarity. Something to keep in mind if you let your agents select their own headset style.

**500 SET** The old rotary dial telephone deskset. The touchtone version was called a 2500 set.

**56 KBPS** A 64 Kbps digital circuit with 8 Kbps used for signaling. Sometimes called Switched 56, DDS or ADN. Each carrier has its own name for this service. The phone companies are phasing out this service in favor of the ISDN BRI, which has two 64 Kbps circuits and one 16 Kbps packet service.

**5ESS** A digital central office switching system made by AT&T. It is typically used as an "end-office," serving local subscribers.

**64 KBPS** A 64 Kbps circuit. "Clear Channel" is 64 Kbps where entire bandwidth is used. Compare to 56 KBPS.

**700 SERVICE** An "area code" reserved for long distance company use. With some companies, dial 1-700 and a "local" telephone number and your long distance carrier will carry that local call. (Whether they can do this or not depends on state law.) Obviously, there is no reason to use this service unless the long distance carrier charges less than the Bell Operating Company (or other local phone company) for the same call.

**800 PORTABILITY** For inbound call centers May 1, 1993 is a date almost equal in importance to the day AT&T divested of the Bell System. On that day 800 service customers were given ownership of their 800 numbers and allowed to take those numbers with them when the changed long distance carriers. 800 portability also lets you split service on a single phone number between two or more carriers. 800 portability drove down the cost of 800 service and increased its usage, especially among small businesses. It led to a shortage of 800 numbers, which in turn led to the creation of a new toll-free code, 888. These days it should be called "toll-free number" portability, since the concept is valid with all the toll-free prefixes.

**800 SERVICE** Eight-hundred service. A common term for "toll-free" or "called party pays." No longer a valid term, because there are so many toll-free prefixes. Pretty soon you will prove yourself an old-timer by referring to "800 service." The popularity of "800 service" after portability meant a new code or prefix was needed to handle all the requests for new numbers. That new code was 888. The 888 code was quickly depleted and the 877 code was added. The other new codes (866, 855, 844, 833 and 822) will be introduced in that order, as needed.

Will 800 numbers become more prestigious as they begin to signify a business that has been in business for a long time? Will consumers accept the new toll free code or be hopelessly confused by the fact that some companies' numbers are valid in both exchanges while others are valid only in one? Will people who still call it "800 service" seem ridiculously out of date? Time will tell. See TOLL FREE SERVICE.

**877** A recently opened toll-free prefix, which joins 800 and 888. The 877 prefix started in April 1998. The big question is why were the 888 numbers used up so fast? It took 20 years to use up the 800 numbers and just two years to run out of 888 numbers. If the problem isn't solved, we'll quickly rip through 866, 855, 844, 833 and 822, the next toll-free prefixes. See 800 SERVICE and TOLL FREE SERVICE.

**888 SERVICE** As of March 1996, the 888 "area" code is an additional code for toll-free or called-party-pays telephone service. This code is in addition to, not in replacement of, the familiar 800 code. The telecom industry and the FCC promise that 888 service will work exactly like 800 service. See TOLL FREE SERVICE.

**900 MEGAHERTZ** A radio frequency band that actually extends from 902 MHz to 928 MHz. It was designated by the FCC for miscellaneous applications, but is a favorite of cordless telephone manufacturers. This frequency band is supposed to go through walls and other barriers more easily than the frequency band used by the old cordless phones.

**900 SERVICE** A generic name for a pay-per-call service where a caller dials a telephone number with a 900 prefix and pays a premium rate for the call. The call usually provides information, such as weather, technical support, entertainment (read chat lines and phone sex lines), games and sports scores.

The term is not trademarked and all the major carriers use the term.

The premium charge for the call appears on the caller's telephone bill. There is no limit to the amount that can be charged (it's usually a per-minute fee), but federal laws to say that the charges must be stated up front in what the industry calls a "kill message." Federal law also says a telephone company can not terminate service because of failure to pay for a 900 charge.

The reputation of 900 numbers suffered on both ends. First, the business was promoted as a get rich quick scheme, especially when the services first became popular. But there is no evidence that running a 900 number takes any less brains, determination or hard work than any other business, so many people felt ripped off.

Second, some 900 services thrived by charging exorbitant fees for services of questionable value. Some did everything the could to have callers spend as much money as possible while delivering as little product as possible. This included keeping people listening to music-on-hold while they waited for their sex talk. Not the way to drum up repeat business.

With stricter laws and many burned bridges, the 900 service industry is different today. It is being used for pay as you go technical support, and other business uses. Whether it can overcome the stigma of its boom years is yet to be seen.

**9001** ISO 9001 is a rigorous international quality standard covering a company's design, development, production, installation and service procedures. Put together by the ISO (International Standards Organization) in Paris, ISO 9001 compliance is becoming more important when doing business overseas — especially in Europe.

**958** Dial 958 in New York City and a computer run by Bell Atlantic reads you the phone number you're calling from. Very helpful when you've just put in a new line and want to make sure they gave you the right number, or if you've forgotten which jack or other connection goes to which number. Other phone companies have similar services but they have different numbers.

**976** The telephone exchange prefix assigned to pay-per-call services limited to a local area by most regional phone companies. Sort of a local 900 number.

# A

**AA** See AUTOMATED ATTENDANT.

**AABS** A payphone/operator services term. This software feature lets callers place collect and third-number billed calls without speaking to a live operator. Call acceptance is validated using a synthesized operator's voice ("Will you accept a collect call from...") and a digital recording of the caller's voice ("Mary"). Calling card services are automated in a similar way. One publication ("The Operator, Volume III, Number 7, April, 1996) put it this way: "Automated Alternate Billing Systems (AABS) have driven substantial costs out of the network with little or no adverse impact on the service that is delivered."

**ABANDONED CALL** An incoming call answered by your ACD, which is terminated by the person originating the call before it is answered by an agent. Usually the caller hangs up because she feels she has waited long enough. (But there are other reasons.) ACDs generally keep statistics on how long your callers wait before disconnecting and what percentage of your calls end this way. This is valuable information to have when planning service levels or creating message-on-hold announcements. Your service level should aim to have most calls answered within the length of time your average caller hangs up. Your message on hold should take into account the amount of time your callers spend waiting for you. Your message should be longer than the average person waits. Using a message on hold can reduce the amount of abandoned calls you have.

There is another, less common use of the term abandoned call that has to do with outbound calling. Sometimes when a predictive dialer is used, it places more calls into the network than there are agents available to handle them. (See PREDICTIVE DIALING for an explanation of how this works.) When there are too many connections and not enough reps, the dialer may hang up on some of those people it called; this is sometimes referred to as an abandoned call. It's also known as a NUISANCE CALL.

**ABANDONED CALL COST** The amount of revenue lost because of abandoned calls (the inbound kind). This is calculated based on the number of calls, the percentage abandoning, and your estimate of the revenue per call. It's an impossible number to calculate since many callers do, in fact, call back and place their orders on another later call.

**ABOVE HOLD TIME** An incoming call that is longer than the average call length for a call center or group.

**AC POWER** Alternating current. Phone systems typically run on AC (meaning they plug into an AC wall socket) and need their own dedicated AC power line. This line should be "cleaned" with a power conditioner and voltage regulator. It also should be protected with a surge arrestor. If possible, the phone system should also be backed by a battery-based power pack, called UPS or Uninterruptible Power Supply.

**AC TO DC CONVERTER** An electronic device which converts alternating current (AC) to direct current (DC). Most phone systems, computers and consumer electronic devices (from answering machines to TVs) run on DC. Even though your phone system plugs into an AC wall socket, it runs on DC from an internal AC to DC converter. Hint: it's probably buried in the power supply.

**AC-DC RINGING** A common way of signaling a telephone. An alternating current (AC) rings the phone bell and a direct current (DC) is used to work a relay to stop ringing when the called person answers.

**ACCESS METHOD** A term used in STRUCTURED WIRING (the coordination of wiring plans within a call center). It's the method of "communicating" on the wire. Examples include Ethernet, Token Ring, AppleTalk, and so forth.

**ACCOUNT CODE (VOLUNTARY OR ENFORCED)** A code assigned to a customer, a project, a department or a division. Typically, a person dialing a long distance phone call must enter that code so a computer can bill the cost of that call at the end of the month or designated time period. Many service companies, such as law offices, engineering firms and advertising agencies use account codes to bill their clients.

**ACD** See AUTOMATIC CALL DISTRIBUTOR.

**ACD APPLICATION BRIDGE** The link between an ACD and a database of information resident on a user's data system. It lets the ACD communicate with a data system and gain access to a database of call processing information.

**ACD APPLICATION-BASED CALL ROUTING** In addition to the traditional methods of routing and tracking calls by trunk and agent group, newer ACDs route and track calls by application. An application is a type of call, for example, sales or service. Tracking calls in this manner allows accurately reported calls, especially when they are overflowed to different agent groups.

**ACD CALL BACK MESSAGING** This ACD capability lets callers leave messages for agents rather than wait for a live agent. It helps to balance agent workloads between peak and off-peak hours. In specific applications, it offers callers the option of waiting on hold. A good example is someone who only wishes to receive a catalog. Rather than wait while other people place extensive orders, they leave their name and address as a message for later follow-up by an agent. Some call cen-

6

ter managers find that if callers leave a message to have their call returned later (so they can place an order, perhaps), the call backs are difficult to schedule, and wind up costing the company more money to process than simply adding more agents. A great idea, though. It needs some thoughtfulness to make it work.

**ACD CALLER DIRECTED CALL ROUTING** Sometimes referred to as an auto attendant capability, this ACD function lets callers direct themselves to the appropriate agent group without the an operator. The caller responds to prompts (For sales, press 1; for service, press 2) and is automatically routed to the designated agent group.

**ACD CONDITIONAL ROUTING** The ability of an ACD to monitor various parameters within the system and call center and to intelligently route calls based on that information. Parameters include volume levels of calls in queue, the number of agents available in designated overflow agent groups, or the length of the longest call. Calls are routed on a conditional basis. "If the number of calls in queue for agent group #1 exceeds 25 and there are at least 4 agents available in agent group #2, then route the call to agent group #2."

**ACD DATA DIRECTED CALL ROUTING** A feature that lets an ACD automatically process calls based on data provided by a database of information resident in a separate data system. For example, a caller inputs an account number via touch tone phone. The number is sent to a data system holding a database of information on customers. The number is identified, validated and the call is distributed automatically based on the specific account type (VIP vs. regular business subscriber, as an example).

**ACD-DN** Automatic Call Distributor - Directory Number. A Nortel term. The queue where incoming calls wait until they are answered. Calls are answered in the order in which they entered the queue.

**ACD INTELLIGENT CALL PROCESSING** The ability of ACDs to intelligently route calls based on information provided by the caller, a caller information database and system parameters within the ACD such as call volumes within agent groups and number of agents available.

**ACIS** Automatic Customer/Caller Identification. This is a feature of many ACD systems. ACIS allows the capture of incoming network identification digits such as DID or DNIS and interprets them to identify the call type or caller. With greater information, such as ANI, this data can identify a calling subscriber number. This is also possible by employing a voice response device to request an inbound caller to identify themselves with a unique code. This could be a phone number, a subscriber number or some other identifying factor. This data can be used to route the call, inform the agent of the call type and even pre-stage the first data screen associated with this call type automatically. See ANI.

7

**ACM** See AUTOMATIC CALL MANAGER.

**ACOUSTIC COUPLER** A connection between a telephone handset and a computer or computer-like device. The handset transmitter and receiver fit into to cups. They are hardly used anymore, except for in older TDDs. Acoustic couplers were once used for laptop users to make a modem connection from a coin phone or where the phone jack is unavailable. It's hard to imagine anyone doing this anymore, but it is still possible.

**ACS** See AUTOMATIC CALL SEQUENCER.

**ACTIVITY CODES** In order to get an accurate picture of what an agent does with his or her time during and after a call, the software applications that the agent uses will often segment their activities by type. Sometimes this happens automatically, when a call is ended, for example, the system notes that the agent moves from one state to another. Other times, the agent has to enter a code into the system, telling it what kind of work he or she is doing. These activity codes are used for reporting on individual and group performance.

**ADA** 1. Average delay to abandon. An ACD statistic that tells you, on average, how long callers are waiting in queue before they hang up. 2. Also, the Americans With Disabilities Act, a Federal law. This law is important to call centers because it says the same services must be available to all customers whether they have a handicap or not. In other words, if your hearing callers can get 24-hour service through your IVR system, your non-hearing customers must also be able to get 24-hour service, either through a TDD-compatible IVR system, or some other method.

**ADAD** Automatic Dialing and Announcing Device. Device which automatically places calls and connects them to a recording or agent. A Canadian term for an automatic dialer.

**ADAPTIVE LEARNING** A method of finding solutions to customer problems implemented in a help desk system. First used by a company called Software Artistry (they have long since been subsumed into IBM, which bought the company in the late 1990s). The concept combines aspects of full-text searching with the self-learning qualities of a neural network. Adaptive Learning "learns" by strengthening the links between keywords and their knowledge base content. In other words, if it finds a link that results in success (a customer problem solved) it is more likely to look to that link in the future, assigning it greater weight.

**ADAPTIVE PULSE CODE MODULATION** APCM. A method of encoding analog voice signals into digital signals that reduces the number of bits required, as compared to the more popular pulse code modulation (PCM). PCM remains more popular, because the electronics required for it are less expensive than those required for adaptive pulse code modulation.

**ADH** Average delay to handle. An ACD statistic that tells you the average amount of time a caller waits before being connected to an agent.

**ADHERENCE** As in "adherence to the staffing schedule." Are your call center agents sticking to the schedule worked out by you and your pencil or your scheduling software? If they are, they are "in adherence." If they are not, they are "out of adherence." There are several reasons for your staff to be out of adherence. Obviously, someone might be out sick. Someone might be late getting back from lunch or break. Also, someone may late going to lunch or break, because a call ran long, or because they would rather take a break later and haven't thought through the staffing consequences of messing with the schedule. One of the hottest features of call center management software and scheduling software is an adherence feature or module, which keeps track of all of this for you.

**ADHERENCE MONITORING** Adherence monitoring means comparing real-time data coming out your ACD with your forecasts, especially staffing levels. Knowing the difference between the forecast and the reality helps you forecast better in the future. There are software packages that will alert you if your staff is out of adherence with their schedule. To do this without automated help is nearly impossible, especially in a larger center. The software keeps track of who's on the phone, who's not, who's late going to break, and who's late coming back from lunch. It can add valuable insight to what is going on when your queue statistics start to go sour. Increasingly, adherence monitoring is a key feature of software that manages information across call center networks, in linked "virtual" centers, and in centers that use skills-based routing criteria.

**ADJUNCT PROCESSOR** A computer outside a telephone switching system that gives the switch commands. An adjunct processor might be a database of customers and their recent buying activities. If the database shows that a customer lives in Indiana, the call from the customer might be switched to the group of agents handling Indiana customers.

**ADPCM** Adaptive Differential Pulse Code Modulation. A speech coding method which calculates the difference between two consecutive speech samples in standard PCM coded telecom voice signals. It allows encoding of voice signals in half the space PCM takes.

**ADRMP** Pronounced "Add-rump." See AUTOMATIC DIALING RECORDED MESSAGE PLAYER.

**ADSI** Analog Display Services Interface. ADSI is a standard defining a protocol on the flow of information between something (a switch, a server, a voice mail system, a service bureau) and a subscriber's telephone, PC, data terminal or other communicating device with a screen. This protocol adds text to your standard voice telephone call. It can, for example, send a screen-equipped phone a complete menu for

**9**

an IVR system, rather than announcing the choices and hoping the user remembers what to do. There are ADSI-compatible phones and ADSI-compatible IVR systems on the market. The drawback is the phones are more expensive than average consumer phones and consumers won't buy them until a killer app drives them to do it. Is ADSI dead? With the rise of the Internet and Internet telephony, there is certainly much less use for it.

**ADSL** Asynchronous Digital Subscriber Line. A method of sending a lot of data over regular (twisted pair copper) telephone lines. It is capable of sending much more information from the big, wide world to your home or office than it is sending things from you out into the world — hence, the "asynchronous" part. When you think about it, this is the way most data transactions work. You download a big, juicy, full-color Web site, but only send them a few clicks of the mouse to navigate it.

Touted as a replacement to ISDN, ADSL has techno-geeks excited because it will let you watch digitized movies over your phone line, or surf the Web at warp speed. Call center folks should be excited because it is just what they need to send their agents home to work. ADSL lets your agents download lots of customer data, product data and applications quickly, and return the much smaller order data or changes. Don't bet the store on ADSL. Like ISDN, it will probably be obsolete before it is widespread.

**ADVISORY TONES** Signals such as dial tone, busy, ringing, fast-busy, call-waiting and camp-on that your telephone system uses to tell you that something is happening or about to happen in the processing of the call.

**AEMIS** Automatic Electronic Management Information System. This was the first computerized UCD/ACD reporting system introduced by AT&T for CO UCD (Central Office Uniform Call Distribution system). Just a little fact for you history buffs. Just about the only place you'll see this term (or the old-style equipment it refers to) is in the museum of obsolete technology.

**AFTER-CALL WORK** The tasks done by an agent after the customer call has ended. This work might be completing an order form or complaint form and sending it to the appropriate department. It might be fulfillment — actually addressing the catalog requested by the caller and sending it to the mail room. It might be conferring with a company expert to check a fact. After-call work is usually done immediately after the call is disconnected. When there are high call volumes, sometimes this work is postponed until an off-peak period. Some predictive dialers and ACDs build after-call work time into their routing algorithms. With this feature the next call does not arrive until after the average time required for after-call work. Also known as AFTER-CALL WRAP-UP.

**AFTER-CALL WRAP-UP** The time an employee spends completing a transaction after the call has been disconnected. Sometimes it's a few seconds. Sometimes it can be minutes. It depends on what is required. See AFTER-CALL WORK.

**AGC** Automatic Gain Control. AGC is an electronic circuit in headsets, tape recorders, speakerphones, and other voice devices. It is used to maintain volume. If the control of the volume is not passed to the party speaking, AGC could amplify the room noise or circuit static. The alternative is manual gain control, where you make every adjustment by hand as the sounds vary. This is not a real option for headsets.

**AGENT** A general term for someone who handles telephone calls in a call center. Other common names for the same job, include, but are not limited to: operator, attendant, representative, customer service representative, CSR, customer support representative, telephone sales representative, technical support representative, TSR, inside salesperson, telephone salesperson and telemarketer.

Some call centers spend much time and effort coming up with a creative job title for their agents. This can do much good, as long as all keep in mind that technology vendors are not shooting for job title creativity, they just use a term they think everyone will understand: agent. Even if your agents are highly trained registered nurses or vain stockbrokers, they should know when the manual or technology label says "agent," that means them.

In the world of computer programming, an agent is a bit of programming that carries out tedious computer tasks, like scanning databanks for relevant information, scheduling meetings or cleaning up e-mail in-boxes.

**AGENT CALLBACK BUTTON** One of your customers is looking at your Web site and would like to order, but doesn't trust the Web to transmit his credit card number safely. Or, a customer wants to know the difference between the cerulean blue bath tile and the cyan blue bath tile and can't tell from the way her computer screen displays your Web page. When you designed your Web site, you thought your customers might need to talk to an agent while surfing, so you put in an agent callback button. The customer clicks his or her mouse and solves the problem.

Most simply, the agent callback button is just the thing that appears on your Web page, but if there were no system linked to the button it wouldn't be much good. Usually after the customer clicks on the button, she gets a little form to fill out with her phone number and perhaps her name and the best time to call back. The systems routes these calls to call center agents in the way you choose. For example, the call could go to the next available agent. That agent would get a screen showing where the customer was on your Web site and a button to push to complete the call. Or, the system might automatically dial the call and deliver the call and the Web page to the agent when the person is on the line.

**AGENT ID** A term used by Rockwell for its now defunct, but ever-popular in its day, Galaxy ACD. An agent ID is a numeric identifier used to maintain Agent Performance logging for an individual agent on the Galaxy. Agent IDs may be 1 to 8 digits in length, based on system parameters.

**AGENT LOGON/LOGOFF** The procedure for alerting the ACD to an agent's availability. Agents logon when they begin their shifts and logoff when they end them. On some systems agents hit a single feature key to logon or logoff. On others, they have to punch in a code.

**AGENT PERFORMANCE REPORT** An ACD report that shows the statistics for each person that has logged off since the previous report. This information includes percent of idle time, busy, or unavailable.

**AGENT SIGN ON/SIGN OFF** An ACD feature which lets any agent occupy any position in the ACD without losing his or her personal identity. This is accomplished by having the agent log in at a position using a personal ID. Statistics are collected and consolidated about this agent and calls are routed to this agent no matter where he sits or how many positions he or she occupies at one time.

**AGGREGATOR** A type of long distance service seller. An aggregator signs up for a long distance service's multi-location toll-free or outbound service then resells the service to other businesses. These other businesses are now the other locations for the service. If you sign up with an aggregator you are still an AT&T, MCI WorldCom (or whatever) customer.

**AHT** Average handling time. An ACD statistic that tells you how long, on average, an agent spends on each call. ACDs calculate this differently. Some include after-call work time, some don't.

**AHT DISTRIBUTION** Average Handle Time Distribution. A set of factors for each day of the week that defines the typical distribution of average handle times throughout the day. Each factor measures how far AHT in the half or quarter hour deviates from the AHT for day as a whole.

**AI** See ARTIFICIAL INTELLIGENCE.

**AIN** Advanced Intelligent Network. A term promoted by the company formerly known as BellCore, as well as the Regional Bell Operating Companies, AT&T and virtually every other phone company to describe their networks for the future. While every phone company has a different interpretation of what their AIN is, there seems to be two consistent threads. First, the network can change the routing of calls within it from moment to moment based on some criteria other than the normal, old-time criteria of simply finding a path through the network for the call. Second, the originator or the ultimate receiver of the call can somehow inject intelligence into the network and affect the flow of his call (either outbound or inbound). Initial AIN services tend to be focused on inbound toll-free calls.

**ALARM** An indication of trouble that is or may become service-affecting. The idea behind an alarm is to attract the attention of someone who can fix the problem.

Alarms can be a sound, a light, a message on a readerboard or a computer terminal, or even a page.

**ALL TRUNKS BUSY** ATB. When a user tries to make an outside call through a telephone system and receives a "fast" busy signal (twice as many signals as a normal busy in the same amount of time), he is usually experiencing the joy of All Trunks Busy. No trunks are available to handle that call. The trunks are all being used at that time for other calls or are out of service. These days, many long distance companies are replacing a "fast" busy signal with a recording that might say something like, "I'm sorry. All circuits are busy. Please try your call later."

**ALPHANUMERIC** A set of characters that contains both letters and numbers — either individually or in combination. Numeric is 12345. Alphabetic is ABCDEF; Alphanumeric is 1A4F6HH8.

**ALPHANUMERIC DISPLAY** In a call center, this usually refers to a display on a phone or agent console that displays the number calling, the number called, trunk number, labels for softkeys, ACD statistics (number of calls in queue, longest call waiting), and possibly information about the caller or the name of the queue. The display may be an LED or light emitting diode, but usually, it's an LCD.

**ALTERNATE ROUTING** A phone system feature usually associated with PBXs, that lets the system send calls over alternate phone lines because of congestion of the lines the calls would normally be sent over. Similar to, but different from least cost routing.

**AMERICAN TELEMARKETING ASSOCIATION** The former name of the American Teleservices Association, before "telemarketing" became a really unpalatable word.

**AMERICAN TELESERVICES ASSOCIATION** A trade and lobbying organization that works on behalf of the telemarketing industry, both on the national level and in individual states. It is headquartered in Los Angeles. In our opinion, every legitimate telemarketing (outbound telephone sales) organization should belong to the ATA. They do good work for the industry.

**AMERICAN TELEPHONE AND TELEGRAPH** See AT&T.

**AMIS** See AUDIO MESSAGING INTERCHANGE SPECIFICATION.

**ANI** Automatic Number Identification. The digits that arrive at the same time as a telephone call that tell you the telephone number of the person calling you. Once this was the favored term for call centers, since ANI was provided by long distance carriers and CLId was the local telephone offering used by people in their homes. The two services use different standards. CLId delivers the digits between the first and second ring. ANI uses a variety of methods: touchtone digits inside the phone call or in a digital form on the same circuit or on a separate circuit. It may arrive over the D channel of an ISDN PRI circuit or on a dedicated single line before the first ring. In Canada the standards were always the same for both local and long distance services.

Then there was a (US) federal mandate for long distance carriers to provide CLId, and ANI was on its way out. But ANI is still provided by long distance carriers. You generally need dedicated access to your carrier's POP to get ANI.

ANI has big benefits for call centers. By gathering the digits sent and doing a database lookup, your agents can receive a screen of information on the caller along with the voice call. Centers report this saves them up to 30 seconds per call, since the agent doesn't have to ask for and enter a name or account number, then wait for the database to respond during the call. Those 30 seconds per call have a significant impact on staffing needs and telephone service charges. ANI can also serve as a security ID for various applications, not the least of which is the local pizza place which uses CLId to screen out crank orders.

The rectangle represents a digital announcement system and its role in a typical inbound call flow pattern. The music- or message-on-hold announcer is represented by the circle. The on-hold announcer is sometimes a slightly different technology and sometimes it is a digital announcer that has been programmed for on-hold play.

**ANNOUNCEMENT SYSTEMS** An "announcement system" and an "announcer" are the same thing. An announcement system is part of the voice processing family. It's a device that answers a call, delivers a message, but does not record a reply. In some applications, the system then puts the caller on hold (or sends him or her to a queue). In others, it disconnects the call. Some of the most common applications for announcers include handling ACD overflow calls; giving general information before handing the call to a live receptionist; giving public information announcements (often used with 800 and 900 programs); and playing music or promotional messages while callers are on hold. Announcers can use tapes or digital technology to store and replay the messages. See DIGITAL ANNOUNCER.

**ANNOUNCERS** See ANNOUNCEMENT SYSTEMS.

**ANSI** American National Standards Institute. An organization that sets the US standards for everything from bicycle helmets to communications protocols. It is not a government agency, and its standards are for voluntary use. It also represents the US in the international standards-setting organizations. A particular ANSI standard can be a good benchmark for comparing products — especially electrical products and power supplies.

**ANSWERING MACHINE DETECTION** A feature of predictive dialers. Dialers need to make an instant decision when the call is answered: send the call to the agent, or not. If it hears a voice, chances are that the call will go to an agent. The ability to detect the difference between an answering machine and a real person can make a sharp difference in the productivity of a dialing system (and the humans who use it). Dialers have apparently gotten quite good at detecting machines.

**ANTICIPATORY DIALING** An automated outbound dialing system similar to predictive dialing. In this mode the dialing algorithm is tied to statistics for an individual agent, rather than a larger group. It "anticipates" when an agent will be off the line and ready for a call. To have a call ready for that agent immediately, it dials a number while the agent is still on the last call. See PREDICTIVE DIALING.

**AOT** Average out time. An ACD statistic. The average length of outgoing calls placed by agents.

**APCM** See ADAPTIVE PULSE CODE MODULATION.

**APM** Average Positions Manned, the average number of ACD positions manned during the reporting period for a particular group.

**APP GEN** See APPLICATION GENERATOR.

**APPLICATION BRIDGE** A term from the early days of CTI. It was Aspect's ACD to host computer link.

**APPLICATION GENERATOR** A programming tool that assists in the development of a voice application, like an IVR script. The app gen hides the ugly programming details from the user, letting the user sketch out (and test) a call flow diagram with graphics, icons and other visual tools. By taking the programming out of the hands of programmers and putting it into the hands of the person who knows exactly what he or she wants from a voice application, it saves time and money.

APR Average positions required. How many agents are needed in a call center or ACD group to meet a service level.

**AREA CODE EXPANSION** In January 1995 BellCore increased the kind of digits that could appear as the second number in the three-digit area code. Some manufacturers of phone equipment, Rockwell, for example, call this event "Area Code Expansion." The problem with some older telecom switches is they were unprepared for the new selection of digits. They reject any area code that does not have zero or one as the central digit. What Rockwell, and other manufacturers, have done is program their switches to accommodate all future changes to the area code.

**AREA CODE RESTRICTION** The ability of the telephone equipment (or its ancillary devices) to selectively deny calls to specific (but not all) area codes. Area code restriction is often confused with toll call restriction, which is not selective at all. This telephone switch feature is most helpful to general businesses or informal call centers. In formal inbound and outbound call centers agents are mostly passive recipients of delivered calls. Restricting them from dialing certain area codes could stem some abuse, but there are probably better ways to do this.

**ARTIFICIAL INTELLIGENCE** A computer software quality that allows the program, not merely to blindly process information, but to put together a chain of logical ideas, draw conclusions and otherwise mimic higher human thought. Call center people will bump into artificial intelligence in certain help desk software programs, especially "problem solving" programs. In these programs, references to "expert systems" and "knowledge base systems" is a clue that artificial intelligence is at work.

**ARTS** Audio Real Time Status. A Rockwell ACD term. This Spectrum ACD feature lets you enter a password from any touch-tone telephone and get real-time ACD statistics, such as average speed of answer, number of calls in queue and activity by agent group. A nice feature for managing from a remote location or when you find yourself fretting on your vacation.

**ARU** See AUDIO RESPONSE UNIT.

**AS/400** A mid-range mini-computer from IBM that was once very popular in call centers. When you hear the term "legacy system" said with a groan, the person may be talking about a call center still using an AS/400. Now a museum piece.

**ASA** Average speed of answer. An ACD statistic. How long the average caller waits on hold before his or her call is answered by an agent. This is an important measure of service quality, and in many call centers it is THE measure used to provide an idea of service quality at any time.

**ASCAP** American Society of Composers, Authors and Publishers. Believe it or not, this term is in this Dictionary for one reason only: Music On Hold. If you play music on hold, you must know what ASCAP is. It is one of two major organizations that protect the copyrights of those in the music industry. (The other is BMI.) What's important for you to know is that all recorded music (even when it is played on the radio) is protected by copyright, and no one can play the music without paying for it. ASCAP serves as a watch dog for musicians' rights, collecting fees for the right to play songs and fines from those who play songs without first paying the fee.

**ASCII** American Standard Code for Information Interchange. A seven-bit code used as a standard for the exchange of data among communications devices. It has become the de facto standard format for text files, as it can be read by software programs from different vendors.

**ASPECT COMMUNICATIONS** Evidence of the sea-change CRM has made on the call center industry. Once a leading maker of ACDs, it is now a maker of "customer relationship portals." It is still located in San Jose, CA though.

**ASR** Automatic Speech Recognition. See SPEECH RECOGNITION.

**ASSIGNMENT** The process of assigning individual employees to specific schedules in a master file or daily workfile. Master file assignment can be done either manually or automatically (based on employee schedule preference and seniority). In call centers, this assignment is likely to be to a particular ACD group, split or skill group.

**ASTERISK LAW** A state law, first passed in Florida, which allows consumers to designate that they do not want to receive telemarketing calls by having an asterisk appear next to their name in the telephone directory. There is a fine imposed on companies that call people whose names are marked by an asterisk. Oregon has a similar law.

**AT HOME AGENT** See REMOTE AGENT.

**ATA** See AMERICAN TELESERVICES ASSOCIATION.

**AT&T** An extremely large company that provides telecommunications services including long-distance service, wireless service, Internet access, consulting services and credit cards. It was once part of the Bell System, which had a monopoly on all telephone technology and services in the United States.

Let's hear the whole story from Sheldon Hochheiser, AT&T Corporate Historian:

"AT&T was incorporated on March 3, 1885, in New York as a wholly owned subsidiary of the American Bell Telephone Company. Its original purpose was to manage, and expand the burgeoning toll, or long distance, business of American Bell and its licensees. It continued as the 'long-distance company' until December 30, 1899, when in a corporate reorganization, it assumed the business and property of American Bell and became the parent company of the Bell System.

"Until divestiture, January 1, 1984, AT&T was the parent company of the Bell System, the regulated enterprise that formerly provided the bulk of telecommunications in the United States. From 1984 until 1996, AT&T was an integrated provider of communications services and products, network equipment and computer systems.

"On September 20, 1995, AT&T announced that it would be splitting into three companies over the subsequent fifteen months. These companies are: today's AT&T, which provides communication services; Lucent Technologies, a systems and technology company, which provides communications products; and NCR Corp., in the computer business."

**ATB** See ALL TRUNKS BUSY.

**ATHT** Average trunk hold time. An ACD statistic. The average amount of time a trunk is in use.

**ATIS** Alliance for Telecommunications Industry Solutions. A key player in the regulation of toll-free (800, 888 and 877) services. The FCC is relying more on input from the industry to make new rules and standards. ATIS is an umbrella organization over many of those industry committees and forums. Under their aegis falls: the Industry Numbering Committee (INC), the Ordering and Billing Forum/Service Management System/800 Number Administration Committee (OBF/SNAC) forum and the Network Operations Forum (NOF). Not to be confused with the Roswell, Georgia-based IVR vendor also called ATIS.

**ATM** No, not an ATM. That's just an automatic teller machine. ATM also stands for Asynchronous Transfer Mode. It's a very high speed, high bandwidth telecom transmission technology. It uses multiplexing and packet-like switching. In ATM transmission, the usable capacity is segmented into fixed-size cells, each with header and information fields. They can be allocated to services on demand. The CCITT (the international standards organization) is betting on ATM's many benefits for the future broadband network. It's a hot technology.

**ATT** Average talk time. An ACD statistic. The average amount of time the agent spends talking to the caller. Usually timed from when the call arrives at the agent station to the time it is released by the agent. Also, an abbreviation for attendant or American Telephone and Telegraph.

**ATTEMPT** Trying to make a telephone call. Also defined as a call offered to a telecommunications system (such as an ACD), regardless of whether it is completed or not.

**ATTENDANT** The person who works the console of a PBX. Commonly known as the company operator or receptionist. Telecom people use the term "attendant." PBX telephone systems generally don't allow calls to be dialed directly to an extension. Incoming calls must be routed to the correct extension by the attendant. If you're thinking, "Wow, that's boring. That job could probably be automated." It has, with a technology called "automated attendant." Call center managers should note this term, because some die-hard telecom people insist on calling call center agents "attendants." See AUTOMATED ATTENDANT.

**ATTENDANT BUSY LAMP FIELD** Lamps, lights or LEDs that show whether a PBX or key system extension is in use or not. So called because the station set (telephone) with this device on it often appears on the attendant's console.

**ATTENDANT CONSOLE** The specialized telephone set used by a PBX attendant. Also known as the attendant station or operator console. The telephone itself is much larger than the average business telephone system telephone and usually has lots of fancy lights and buttons. On some systems the attendant console looks a lot like the agent's ACD station set, but as the attendant console is primarily a routing device, and the ACD station set is primarily an answering device, they perform very different functions.

**AUDIO MESSAGING INTERCHANGE SPECIFICATION** AMIS. A series of standards that allows messaging systems made by different vendors to exchange voice messages. It does not describe the user interface to a voice messaging system, specify how to implement AMIS in a particular systems or limit the features a vendor may implement. There are analog and digital versions of the specifications.

**AUDIO RESPONSE UNIT** A device which translates computer output into spoken voice. Let's say you dial a computer and it said "If you want the weather in Chicago, push 123, then it would give you the weather. But that weather would be "spoken" by an audio response unit. Here's a slightly more technical explanation" An audio response unit is a device that provides synthesized voice responses to dual-tone multi-frequency signaling input. These devices process calls based on the caller's input, information received from a host data base, and information carried with the incoming call (e.g., time of day). ARUs are used to increase the number of information calls handled and to provide consistent quality in information retrieval. See also AUDIOTEXT and INTERACTIVE VOICE RESPONSE.

**AUDIOTEXT** A voice processing system that presents a caller with a menu of choices, which are selected by pushing a button on a touch-tone telephone, then plays a recorded announcement for that menu choice. Some people use this term to include interactive voice response (IVR) systems, but we prefer to reserve "audiotext" for a stand-alone system that does not interact with a computer database. Audiotext also does not route calls based on menu selection. That would be called an automated attendant. With audiotext, each caller who selects the same menu choice hears the same message. With IVR the message is customized based on information from an external computer database.

Audiotext is very useful when agents in an informal call center are spending all their time answering repetitive questions. (It can also be helpful in a formal call center.) Examples of great and popular uses for audiotext: a schedule of movies and show times for a movie theater; basic corporate information such as mailing address, directions, fax and e-mail numbers; program information for museums, galleries, parks and the like; top 10 technical tips.

**AUDITORY PATTERN RECOGNITION** A fancy way of saying the ability to recognize spoken words.

**AUTO ATTENDANT** See AUTOMATED ATTENDANT.

**AUTODIALING** Dialing a telephone number automatically. "Autodialing" is a general term that describes a host of dialing techniques that range from preview dialing to predictive dialing. See AUTOMATIC DIALER.

**AUTO DIALER** See AUTOMATIC DIALER.

**AUTO FAX TONE** Also called CNG, or Calling Tone. The sound produced by virtually all Group 3 fax machines when they dial another fax machine.

**AUTOMATED ALTERNATE BILLING SYSTEMS** A payphone/operator services term. See AABS.

**AUTOMATED ATTENDANT** A voice processing device that answers calls with a digital recording, then lets callers route themselves to the person or department they want by entering the appropriate extension on their telephone's touchtone keypad. This device is usually connected to a PBX, which because of its basic nature, does not connect a caller directly to a particular extension (a few do, with a special feature). In a very real way, this device automates the PBX attendant's function. Usually the greeting will say, "Thank you for calling Our Company. If you know the extension of the person you wish to speak to, dial it now. For the accounting department press 111, For sales, press 222... Dial zero for the operator, or stay on the line and your call will be answered shortly."

An automated attendant can also be attached to a voice mail system, and there is no reason why it can't help direct calls in your PBX-based ACD.

**AUTOMATED VOICE RESPONSE SYSTEM** AVRS. A device that automatically

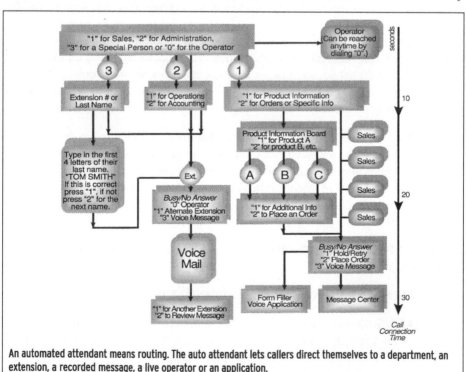

An automated attendant means routing. The auto attendant lets callers direct themselves to a department, an extension, a recorded message, a live operator or an application.

answer calls. It may simply play a message that says the call is in queue and will be answered soon, or it may give the caller further choices through touch-tones or even voice commands. See VRU.

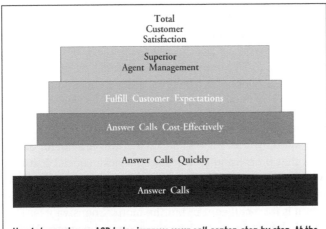

Total
Customer
Satisfaction

Superior
Agent Management

Fulfill Customer Expectations

Answer Calls Cost-Effectively

Answer Calls Quickly

Answer Calls

Here's how using an ACD helps improve your call center, step by step. At the bottom of the pyramid are the most basic goals accomplished with the most basic technologies. Each step represents both a level of service sophistication and technological sophistication.

**AUTOMATIC CALL DISTRIBUTOR** ACD. An ACD answers a call and puts the call in a pre-specified order in a line of waiting calls. On the simplest level, it makes sure the first call to arrive is the first call answered. It delivers calls to agents in a pre-specified order. It delivers the call to the agent who has been free (or idle) the longest or to the next agent that becomes available in a call center. It also provides the means to specify the many possible variations in the order of calls and agents. Last but not least, it provides detailed reports on every aspect of the call transaction, including how many calls were connected to the system, how many calls reached agent, how long the longest call waited for an agent, the average length of each call and many more.

Once the term "ACD" meant a very specific type of telephone switch. It was a switch with highly specialized features and particularly robust call processing capabilities that served at least 100 stations (or extensions). It was purchased mostly by airlines for their reservations centers and large catalogs for their order centers. Companies with less specialized needs bought different technologies that didn't offer the same specialized features. Today true ACD functionality is found in telephone switches that range widely in size and sophistication.

Today there are PC-based ACDs, key systems with ACD functions, key systems that integrate with a computer and software to create a full-featured ACD, PBXs with ACD functions, server-based ACDs, PBXs with ACD functions that are so sophisticated they compete with stand-alone ACD systems, stand-alone ACDs that serve centers with less than 30 agents, traditional stand-alone ACDs (don't misunderstand, these usually the most sophisticated), ACDs that integrate with other call center technologies, and nationwide networks of ACDs that act as a single switch.

There is simply no technology more suited to routing a large number of inbound calls to a large number of people than an ACD. Using an ACD assures your callers are answered as quickly as possible. It can provide special service for special customers. ACDs are capable of handling calls at a rate and volume far beyond human capabilities, and in fact, beyond the capabilities of other telecom switches. It provides a huge amount of call processing horsepower. Using an ACD assures your human resources are used as effectively as possible. It even lets you create your own definition of effectiveness. An ACD gives you the resources to manage the many parts of your call center, from telephone trunks to agent stations to calls and callers to your agents and staff.

If ACDs are so great, why does anyone bother with a general business telephone system (such as a PBX or key system)? The ACD's special features are not for every business. Most important, all that processing power and all those special features come at a price. Expect to spend more "per seat" on an ACD than you would for a PBX or key system without sophisticated ACD features. The more robust and feature-rich the ACD, the better it is for a call center, but the more expensive it is. The processing power alone quickly becomes overkill for the average business. (We've seen calculations that say 30 ACD positions use the same processing power as 100 PBX positions.) The average ACD (although certainly not the cutting-edge ACDs) lacks basic business telecom features. They are sacrificed to give more processing power to the tasks at hand. Making an outbound call is a special event on a stereotypical ACD. (To confuse matters there are ACDs that have predictive dialing functions.) Simple general business telephone functions like picking up the phone and dialing someone in another department can be complicated on an ACD.

The plain-vanilla ACD described in the opening paragraph of this definition isn't too technologically impressive these days. (Although it certainly was over 20 years ago when Rockwell first introduced the technology.) Let's follow a call to explore some of the features you can expect to find these days. First, your call will be greeted by an announcement of some kind. The announcement may simply tell you all agents are busy, please wait. Through integration with a reporting system, it may tell you how long the wait will be. It may ask you to enter your telephone number, account number or ID. The ACD will use this to do a database lookup and present information about you (usually on an integrated computer system) along with the call (SCREEN POP). The announcement may offer you the option of trying to solve your problem through an IVR system, without losing your place in the telephone queue.

The ACD may also gather information about you through automatic number identification (ANI). It may route your call based on the toll-free number you dialed. A single call center may answer hundreds of telephone numbers. Collecting the dialed number identification service (DNIS) info, the ACD directs your call to the correct agent group (customer service, toasters) or even pinpoints a particular agent that can best help you (SKILLS-BASED ROUTING). The ACD's routing scheme can be configured to distribute your call based on a huge range of criteria, including time of day, day of week, volume of calls in the center, number of agents available and many more. ACDs can put VIPs in a special queue, transfer a call on an order line to the collections department based on the customer ID, send a caller to an agent who speaks her language or send a caller to an agent or group which specializes in the product she needs help with.

While the call is processed, the ACD can provide a supervisor with real-time information about calls in the center, a group or the status of a single agent. It allows supervisors to listen in on calls to evaluate agents, or to join in to assist an agent who is having trouble. Call center statistics displayed on an agent's telephone or on a READERBOARD help the agent manage her own time. Integration with a computer system offers a variety of special features that lace the agents' service software with the telephone system.

After the call, the ACD can automatically give the agent a certain amount of time to finish the transaction before feeding the agent the next call. Statistics for the call are added to the reporting system. Through integration with a computer system, all sorts of activities can be generated automatically from the call — from fulfillment to database updates to the scheduling of a callback.

This definition was meant only to give you a glimpse at some of the functions and capabilities of an ACD.

**AUTOMATIC CALL MANAGER** A term used for an integrated inbound call distributor and automated outdialing system. Telemarketing and collections applications are targets for this type of system.

**AUTOMATIC CUSTOMER/CALLER IDENTIFICATION SYSTEM** See ACIS.

**AUTOMATIC CALLING UNIT** ACU. A device that places a telephone call on behalf of a computer.

**AUTOMATIC CALL SEQUENCER** A device that puts incoming calls in a queue. It performs three functions. 1. It answers an incoming call, gives the caller a message, and puts them on hold. 2. It signals the agent which call on which line to answer next. Usually the call it picks to be next is the one that has been on hold the longest. 3. It provides management information, such as how many abandoned calls there were, how long the longest person was kept on hold, how long the average time spent waiting was.

There are two other queuing devices, uniform call distributors (UCDs) and automatic call distributors (ACDs). The automatic call sequencer differs from the other two in that it has no internal switching mechanism and it does not effect the call in any way. Its only role is to suggest (usually through a flashing light) which call should be answered next.

An automatic call sequencer was designed to put some order to the incoming calls for a busy attendant or a receptionist. It is commonly found on a key system. In other words, it is designed for just one person answering a lot of calls. Call center people, being clever, quickly figured out how to use this function for simple and small inbound applications. The more sophisticated UCDs (found on PBXs) and ACDs (often a technology in itself) are actually designed for multiple agent inbound queuing, and therefore have more features. See ACD and UCD.

**AUTOMATIC DIALER** or **AUTODIALER** A generic term for a host of automated dialing technologies, ranging from a simple click on a number field in a software program and dial, to sophisticated systems capable of handling thousands of numbers, and delivering calls to hundreds of agents without pause. Summarizing the functions of an automatic dialer are tricky, mostly because the vendors in the industry seem to argue about them so much. It is therefore with some trepidation we offer this hierarchy of automatic dialing functions, in order from least sophisticated to most sophisticated.

1. Dials a telephone number that appears in a selected field in a computer software program.

2. Dials through a list of telephone numbers sequentially, when prompted by a keystroke from the calling agent.

3. Dials through a list of telephone numbers sequentially, automatically dialing the next number when the previous call is completed.

4. Dials a number and screens the call for busies, no answers, interrupt messages and answering machines. Delivers only calls completed to live humans to the calling agent.

5. Anticipates when a single agent will be finished with a call, and dials the next number so the call will be ready immediately when the call is completed.

6. Anticipates when the next agent in a group of agents will be available, and dials a number so the next call will be ready immediately.

7. Takes into account the number of busies, no answers, intercept messages and answering machines likely to be encountered and times the dialing of the next call to deliver only a live human to the next available agent.

8. Takes into account that the average time agents spend on calls, as well as the number of busies, no answers, etc., changes based on the list, campaign, time of day, day of week and other factors, and adjusts its speed of dialing accordingly.

This hierarchy hits only the high spots, but it gives you the general idea. See PREDICTIVE DIALING, PREVIEW DIALING, and ANTICIPATORY DIALING.

**AUTOMATIC DIALING RECORDED MESSAGE PLAYER** ADRMP. Pronounced "add-rump." A machine that dials multiple phone numbers and plays a message. It is a very primitive form of automated dialing. One that does not connect the called party to a live person. Never popular with consumers, ADRMPs remained popular with marketers until the Telephone Consumer Protection Act of 1992 eliminated "sequential dialing," calling every single phone number in an exchange, in order, by forbidding calls to emergency lines and cellular phones. Practically, ADRMPs can now only be used with consumer-specific lists, which has gone a long way toward eliminating abuse of the technology. ADRMPs are valuable for reaching large numbers of people in an emergency (to evacuate the area around a nuclear power plant for example), to alert people to upcoming appointments (doctor, furniture delivery), for collections and for certain kinds of customer service follow-up.

**AUTOMATIC GAIN CONTROL** See AGC.

**AUTOMATIC NUMBER IDENTIFICATION** See ANI.

**AVAILABLE** Used to describe an agent between calls. An agent that is ready to receive another call. One of the agent "states" tracked in real-time in some call center management software programs.

**AVAILABILITY** The amount of time an agent or agent group is available to receive calls. The amount of time they are logged-in, at their desks, but not on a call. A statistic tracked by some call center management software programs. A similar term is also used for computers or telephone systems for the time they are turned on and available for processing calls or transactions.

**AVERAGE CALL DURATION** The amount of time the average call lasts. Calculated by dividing the total number of minutes of conservation by the number of conversations.

**AVERAGE CUSTOMER ARRIVAL RATE** Represents the number of entities (humans, packets, calls, etc.) reaching a queuing system in a unit of time. This average is denoted by the Greek letter lambda. One would prefer to know, if possible, the full distribution of the calls arriving.

**AVERAGE DELAY** The delay between the time a call is answered by the ACD and the time it is answered by a person. This typically includes time for an initial recorded announcement plus time spent waiting in queue. Average delay can be used as a rough measure of service quality.

**AVERAGE DELAY TO ABANDON** See ADA.

**AVERAGE DELAY TO HANDLE** See ADH.

**AVERAGE HANDLING (or HANDLE) TIME** See AHT.

**AVERAGE HOLDING TIME** The sum of the lengths (in minutes or seconds) of all phone calls during the busiest hour of the day divided by the number of calls. There are two definitions. The one above refers to average speaking time (it's the more common one). There's a second definition for "average holding time." This refers to how long each call was on hold, and thus not speaking. This second definition is typically found in the automatic call distribution business (ACD). Check before you do your calculations.

**AVERAGE OUT TIME** See AOT.

**AVERAGE POSITIONS MANNED** See APM.

**AVERAGE POSITIONS REQUIRED** See APR.

**AVERAGE SPEED OF ANSWER** See ASA.

**AVERAGE TALK TIME** See ATT.

**AVERAGE TRUNK HOLD TIME** See ATHT.

**AVERAGE WAIT TIME** The length of time a caller must spend on hold before an ACD can find an available agent to take the call. Most ACDs will keep track of this statistic. Obviously, the shorter the time the better, especially if your company is paying for the call (such as toll-free 800, 888 or 877 call). Minute reductions in the average wait time can add up to huge savings in toll-free service usage when multiplied by thousands of calls. If you need to add agents to reduce your average wait time, however, you must balance the additional payroll costs against your toll-free service charges and your abandoned call costs.

**AVERAGE WORK TIME** See AWT.

**AWT** Average work time. An ACD statistic. Sometimes called after-call work time or wrap-up time. Some call center managers push and push for this average to be shorter. Doing this can lead to your agents making an excessive amount of mistakes in their after-call work. After-call work is an important part of the customer transaction. AWT can also stand for AVERAGE WAIT TIME.

**B CHANNEL** An ISDN "bearer" channel. See ISDN.

**BACK HAUL** Back haul is a verb. A communications channel is back hauling when it takes traffic beyond its destination and back. One reason to do this is economics. It's cheaper to send calls that way. Another reason is to accommodate changes in your calling or staffing patterns. You may have an ACD in Omaha and one in Chicago. A call from New York may come into your Omaha ACD, but when it gets there you may discover that there are no agents available to handle the call. So it may now make sense to back haul the call to the Chicago ACD, where an agent is available.

**BACKPLANE** The card inside the ACD on which you attach other cards. It is exactly the same as the motherboard in a PC. In fact, increasingly an ACD is in fact a souped-up server PC, as often as not running UNIX or NT. In that case, the backplane is the motherboard.

**BACKWARD SIGNAL** A signal sent in the direction from the called to the calling station, or from the original communications sink to the original communications source. The backward signal is usually sent in the backward channel and consists of supervisory, acknowledgement or error control elements.

**BAND** (1) Back in the old days, real WATS (Wide Area Telephone Service) had real bands. Today most services offer virtual bands on virtual WATS. Real WATS split the US into six bands of states. A ring of states closest to your state make up band one, the next ring of states after that makes up band two, and so on. Each band was carried on a separate trunk or group of trunks. Your own state was not included since WATS is a long distance service.

All calls within a band are priced at the same per-minute rate. You can order whatever bands you want (one of them, all of them or just some). For more, see WATS.

(2) A group of radio frequencies (for example, citizens' band). The microwave radio part of the electromagnetic spectrum is broken up into bands by usage. There are bands for the government, bands for long distance telephone carriers and bands for commercial use — private, company microwave links.

**BARGE-IN** A supervisory feature of ACDs which permits a monitoring supervisor to cut into a call being handled by a service rep. An optional zip tone announcing the barge-in may be sent to the service rep's headset only, allowing him to put the customer on hold while speaking to the supervisor; otherwise, the supervisor will be heard by both the service rep and the customer.

**BASE AMOUNT** A historical pattern of call volume. What the monthly call volume would be if there were no long-term trend or seasonal fluctuation — in other words, the average number of calls per month.

**BASE LOAD** In trunk forecasting, an amount of telephone traffic measured during a certain defined time.

**BASE SCHEDULES** A fixed set of pre-existing staff schedules that you can use as a starting point in scheduling. New schedules created are in addition to the base schedules.

**BASE STAFF** The minimum number of people, or "bodies in chairs," required to handle the workload in a given period. The actual number of staff required is always greater than base the staff because of human factors such as the need for breaks and time off.

**BASIC RATE INTERFACE BRI.** One of two ISDN interfaces — and the smaller one. BRI gives you two bearer, or B-channels, at 64 kilobits per second and a data, or D-channel, at 16 kilobits per second. The D-channel handles information about the calls themselves. The B-channels handle the actual information content of the call, whether it's a Group 4 fax transmission, a videoconference or multiplexed voice calls. One BRI standard is the "U" interface, which uses two wires. Another BRI standard is the "T" interface which uses four wires.

**BATTERY** All telephone systems work on DC (direct current). DC power is what you use to talk on. Often the DC power is called "talking battery." Your ACD probably plugs into an AC outlet, but that AC power is converted by a built-in power supply to the DC power the phone system needs. All central offices (public exchanges) used rechargeable lead acid batteries to drive them. These batteries perform several functions: 1. They provide the necessary power. 2. They serve as a filter to smooth out fluctuations in the commercial power and remove the "noise" that power often carries. 3. They provide necessary backup power should commercial power stop, as in a "blackout" or should it get very weak, as in a "brownout."

**BATTERY BACKUP** A battery which provides power to your phone system when the main AC power fails especially during blackouts and brownouts. Hospitals, brokerage companies, airlines and hotel reservation services must have battery backup because of the integral importance of their phone systems to their business.

**BEEP TONE** A simple sound sent from the phone switch to the agent's headset alerting him to the fact that a call is about to be connected.

**BELL OPERATING COMPANY** See BOC.

**BELLCORE** Bell Communications Research. An organization that serves the seven

28

Regional Bell Operating Companies (RBOCs) with research and other services of common interest. They also coordinate the communications aspects of national security and emergency preparedness for the federal government. They help create and administer the standards and procedures that let telephone networks run by different, and sometimes competing, companies communicate. They are not actually known as Bellcore any longer; the new name for the company is Telcordia.

**BELOW HOLD TIME** A call that is shorter than the average minimum call length. If many of these calls show up on the same trunk, it may indicate transmission problems on that trunk.

**BENCHMARKING** The process of identifying current levels of performance (both in-house and at select outside examples) and using that data to set achievable performance goals. Performance is not the only thing that needs to be benchmarked; you also need to benchmark customer expectations, and match the two.

**BESPOKE** As in "bespoke system." This is a trendy way to say "custom made" or "made to order" system, as opposed to a shrink-wrapped or off-the-shelf system. The term comes from fashion, where a custom-made suit is a bespoke suit. For your call center technology, this customization must be done by the vendor, a value-added reseller or consultant, not a tailor. Avoid this term if you can. It drips of snobbery.

**BHCA** See BUSY HOUR CALL ATTEMPTS.

**BHCC** See BUSY HOUR CALL COMPLETIONS.

**BIG BANG** A cliche used to describe sweeping deregulation (or other regulatory changes) that opens a particular market to competition. The term has gotten a workout in the past few years in the call center industry. The "big bang" of utility deregulation, especially in California, means that utilities all over the country are beefing up their call centers. The "big bang" of European telecom happened on January 1, 1998, when four new countries opened their telecom industries to competition.

**BILLING INCREMENT** The increments of time in which the phone company (long distance or local) bills. Some services are measured and billed in one minute increments. Others are measured and billed in six or ten second increments. Short billing increments become important to you, as a user, when your average calls are very short — for example, if you're making a lot of very short data calls (say for credit card authorizations). Being billed for a lot of six second calls is a lot cheaper than being billed for a lot of one minute calls.

**BINAURAL** A headset with two earpieces or receivers. Walkman headphones are a good example — even though they are not headsets.

**BLEND** To have outbound and inbound phone calls answered by the same agents. See the next two definitions.

**BLENDED AGENT** A call center staffer who answers both incoming and makes outgoing calls. One advantage is more efficient use of the agent. If an inbound agent is sitting idle, he or she is assigned to make outbound calls. If an outbound agent is idle, he or she is assigned to take some of the overflow calls from the inbound group. Another advantage is variety for the long-term, highly-skilled agent. Handling different types of calls makes the job less like assembly line work.

Many managers feel, however, that inbound and outbound agents bring two entirely different skill sets to the job. An agent who excels and smoothing over a customer complaint may have difficulty closing a sale — and visa versa. Also, many call centers have a hard enough time finding even minimally skilled people to fill their seats. In a high-turnover call center, developing agents with skills to handle both inbound and outbound calls might be difficult.

**BLENDED CALL CENTER** A call center where the telephone switch acts as both an ACD (automatic call distributor) and a predictive dialer, allowing agents to both receive and make large numbers of calls as demand and strategy dictate. There are three technological strategies to achieve this. First, to link a stand-alone ACD and a stand-alone predictive dialer together, perhaps through a computer system using computer-telephone integration. Second, is to buy an ACD with predictive dialing features built in. And third, is to buy a predictive dialer with sophisticated inbound call routing capabilities.

**BLOCKED CALLS** Calls that can't can not be completed. The caller usually hears a busy signal to indicate all of the (local or long distance) telephone carrier's trunks are in use or disabled. See BLOCKING.

**BLOCKED CALLS DELAYED** A variable in queuing theory to describe what happens when the user is held in queue because his call is blocked and he can't complete it instantly.

**BLOCKED CALLS HELD** A variable in queuing theory to describe what happens when the user redials the moment he encounters blockage.

**BLOCKED CALLS RELEASED** A variable in queuing theory to describe what happens when the user, after being blocked, waits a little while before redialing.

**BLOCKING** When a telephone call cannot be completed it is said that the call is "blocked." Blocked calls are different from calls that are not completed because the called number is busy. This is because numbers that are busy are not the fault of the telephone switching and transmission network.

The "Grade of Service" is a measurement of blocking. It varies from almost zero (best, but most expensive case, no calls blocked) to one (worst case, all calls blocked). Grade of Service is written as P.05 (five percent blocking). "Blocking" used to be a technical term but has now become a sales tool especially among telephone switch manufacturers, who increasingly claim their switch to be "non-blocking." This means it will not, they claim, block a call in the switch.

**BMI** Broadcast Music, Inc. One of two major organizations that protect the copyrights of musicians, songwriters and producers. (The other is ASCAP.) For call centers, the important thing to know is that all recorded music (even when it is played on the radio) is protected by a copyright that says no one can play the music without paying for it. BMI serves as a watch dog for musicians' rights, making sure they get paid the money due them. One way they do this is to collect fees from businesses who play music on hold, either from the radio or a recording. Another way is by calling up businesses that they have no fee-payment record from and listening for violations of music copyright through the use of music on hold. The business in violation has to pay a huge fine. BMI then distributes the fees or fines to the rights-holders in the form of royalties.

**BOC** Bell Operating Company. The local Bell telephone company. Federal law now allows BOCs to consolidate and RBOCs to acquire each other, so keeping track of how many there are is tricky. Also see TELCO and RBOC.

**BOOM** The wire-like piece that attaches a headset microphone to the earpiece and holds it in the correct position in front of your mouth.

**BRANCHING SCRIPTS** In telemarketing or sales, the prewritten guide that tells the salesperson what to say is called the "script." It can spell out every word the salesperson says, or it can give a sketchier outline of points to raise, depending on the circumstances. Some contact management and sales software programs let you create a script that accounts for different outcomes, called a branching script. That is, if the prospect raises an objection to price, the seller hits a key and a response comes up. Other objections would lead to different sets of responses. Some programs let you have hundreds of conversational pathways.

**BREAK OPTIMIZATION** The automatic adjustment of break start times for schedules to more closely match staff to workload in each period of the day. A software program with this feature can improve on the originally scheduled break arrangement because it now has information about schedule exceptions, newly added schedules, and additional call volume in AHT (Average Handle Time) history. See BREAK PARAMETERS.

**BREAK PARAMETERS** A group of assumptions you set to govern the placement of breaks in employee scheduling. These are typically:

• Earliest allowable break start time • Latest allowable break start time • Duration of the break • Whether the break is paid or unpaid

**BRI** See BASIC RATE INTERFACE.

**BUCKET SHOP** A shady telemarketing call center. A bucket shop values short term results (read profits) over customer satisfaction, long term stability, and in some cases, legality. Bucket shops tend to be very low tech. (Customer records on index cards stored in shoe boxes. Manual dialing. Basic telephone sets.) The better to pick up and move if hounded by dissatisfied customers or regulatory agencies.

**BURN OUT** A condition, where stress causes agents to be apathetic and lethargic, caused by intensity of calling, lack of variety and poor working conditions. It is particularly associated with outbound cold calling and inbound complaint handling, both of which are stressful for agents if not carefully managed.

**BURN 'EM AND CHURN 'EM** A call center "management" style. In this philosophy, the human needs of the call center agent are ignored. The pressure is high, either for results or the sheer number of calls handled (or both). Staff development is not attempted. When the agents burn out and leave, new agents are hired and the cycle continues.

**BUSINESS PROCESS ANALYSIS** Call centers should not be isolated from the rest of the company. All too often the decisions about how to deal with customers are made in isolation, while customer interactions are occurring in several different places at the same time. By that we mean that a call center will have one set of rules for dealing with a customer who calls, and a marketing department will have another set of rules for dealing with letters and faxes, and maybe there's a webmaster who is sitting on hundreds or thousands of customer e-mails, without any rules for how to respond. That's a mess of inconsistency, and it spells disaster for a company that doesn't collate and manage these rules.

Business process analysis is a way of grabbing hold of all of a company's modes of interacting with customers, coordinating them across the organization, and making them consistent one to another. It used to require an army of consultants and custom-developed software systems to make sure that no matter how a customer came to you, you had all the information neatly formatted about that customer's history, likes, dislikes and value to the company.

Now, it can be done with an emerging software category called "customer relationship management" systems (which grew out of a combination of help desk software, workflow, and telemarketing/call tracking systems). And, still, an army of consultants. But the benefits can be huge: you can know, for example, that a particular customer is worth huge revenues, and should be put to the head of the queue, or answered by an agent with particular skills, or that he sent an e-mail about a problem that still hasn't been resolved. This data is priceless.

On the downside, company-wide rules about customer interaction are often made outside the call center, diminishing the role of call center management in the decision making process. All participants share information, including marketing, product development, support and sales, as well as the IT infrastructure who's data it ultimately is.

**BUSINESS-TO-BUSINESS** Marketing — usually direct — that targets businesses as customers. Business-to-business telephone sales requires a vastly different strategy from business-to-consumer telephone sales. Calls to businesses are usually answered, but not by the major decision makers. Predictive dialers are rarely, if ever, used for business-to-business telephone sales.

**BUSINESS FILE** A database of business phone numbers, addresses, or overlaid details, used for business-to-business marketing.

**BUSY** In use. "Off-hook". There are slow busies and fast busies. Slow busies are when the phone at the other end is busy or off-hook. You hear the buzz 60 times a minute. Fast busies (120 times a minute) occur when the network is congested with too many calls. Your distant party may or may not be busy, but you'll never know because you never got that far.

**BUSY BACK** Telecom talk for busy signal.

**BUSY HOUR** The uninterrupted period of 60 minutes for which the average intensity of traffic is at the maximum. it is the busiest hour of the busiest day of the normal week, excluding holidays, weekends and special event days. Knowing when your center's busy hour occurs (and what the call volume is when it occurs) is vital for staff scheduling, traffic engineering and equipment purchases.

The idea is if you create enough capacity to carry that "busy hour" traffic, you will be able to carry all other traffic. In actuality, one never designs capacity sufficient to carry 100% of the busy hour traffic. That would be too expensive. So, the argument then comes down to, "What percentage of my peak busy or busy hour calls am I prepared to block?" This percentage might be as low as half of one percent or as high as 10%. Typically, it's between 2% and 5%, depending on what business you're in and the cost to you — in lost sales, etc. — of blocking calls.

**BUSY HOUR CALL ATTEMPTS** BHCA. The maximum number of incoming call attempts an ACD handles in a given hour. The attempt is an incoming call recognized by the system. Often confused with busy hour call completions. BHCA can either be an ACD system specification, determined by the manufacturer through engineering theory or experimental testing, or it can be a call center statistic, tracked and reported by the ACD on a periodic basis.

**BUSY HOUR CALL COMPLETIONS** BHCC. The number of trunk seizures an ACD can handle and the number of calls the system can actually process through a normal cycle. BHCC can either be an ACD system specification, determined by the manufacturer through engineering theory or experimental testing, or it can be a call center statistic, tracked and reported by the ACD on a periodic basis. It is similar to busy hour call attempts, but is probably the more important specification to consider when purchasing an ACD. Ask the vendor how the specification was determined (theoretically or experimentally) and see if you can get a BHCC statistic from an actual, large user.

**BUSY HOUR CALLS** The number of calls anticipated during the busy hour.

**BUSY HOUR USAGE PROFILE** The busy hour usage profile identifies how a system will normally be used (i.e., who the users are and what type of transactions they are performing) during the busy hour. This is a particularly important concept in integrated computer/telephone systems, where vastly different actions put different stresses on the system.

**BUSY OUT** To cause a line to return a busy signal to a caller. This is also known as "taking the phone off the hook," although with an ACD, there are more sophisticated ways to do this. In an ACD with trunks that rotary (or hunt) on, sometimes busying out one or more broken trunks helps calls move on to trunks that are still working. This way, someone doesn't end up on your third trunk with endless ringing, while your 4th, 5th and 6th trunks are free. This is also a way to free up agents for after-call work, coffee breaks and other interruptions to receiving calls.

**BUSY SEASON** An annual recurring and reasonably predictable period of peak requirements. For inbound call centers of companies that sell consumer goods (read, "catalogs"), the busy season starts in November and ends sometime in late January or early February, when all the Christmas gifts have been returned. For financial services companies, busy season stretches from the end of the year until April 15. For gardening supply companies, it starts in February and ends in late June or July. (You get the point that it varies from industry to industry.) During these periods, some call centers bring in temporary workers, or have existing staffers work more hours. Many call centers are open longer hours during their busy season than they are the rest of the year. Most call center managers know when their center's busy season by noting their own exhaustion. A better way is to use historical records of call volume, such as those from a call center management software package.

**BUTT SET** A butt set is the device that sort of looks like a telephone handset, that telecom repair people like to have dangle off their belts. It often has alligator clips to latch on to phone lines to make or monitor calls. It's called a butt set because it lets them "butt in" on calls.

After you get some training, a butt set is a valuable piece of test equipment that can help you figure out what's wrong before you call the telephone company.

**BYPASS** An ACD feature that lets you connect agents directly to telephone lines when the ACD fails (or is shut off).

**CABLE TYPE** A term used in STRUCTURED WIRING (the coordination of wiring plans within a call center). This refers to what kinds of wire (cable) you use, for example, coaxial, UTP, STP and fiber. Factors including cost, connectivity and bandwidth are important in determining cable type.

**CALENDAR ROUTING** A method of directing calls according to the day of the week and time of day. See also SOURCE/DESTINATION ROUTING, SKILLS-BASED ROUTING and END-OF-SHIFT ROUTING.

**CALIBRATE** To standardize the scoring of the quality of customer interactions. The process of calibration is most useful in situations where supervisors are monitoring agents. Calibration involves a long process of hashing out the goals of calls and the expectations of agents and supervisors.

The process starts with recording calls, then having a set of monitors discuss the calls and argue about their varied interpretations of how the call was handled. From there, you begin to set more accurate standards.

**CALIBRATION** The process of standardizing call center quality. See CALIBRATE.

**CALL** What is a call? Traditionally, we think of a call as a voice telephony connection between two parties, one the originator and one the receiver.

Nowadays, particularly in call centers, we need to think of a much broader definition of the word "call." Interactions between company and customer can happen in so many ways, in fact, (all of them electronic) that it's hardly useful to think of the call center as merely a place where voice calls terminate.

A call can be defined as an electronic interaction between a company and a customer (actual or potential) for the purpose of exchanging information. It can start on either end (either inbound or outbound, from the center's point of view). And it can be in the form of a traditional voice call, or an IVR hit, an interactive web exchange (where the "caller" fills in a form or is led from page to page by an agent pushing pages down the line), e-mails, live text chat over the internet, a video kiosk connection, or any other electronic connection. Just about the only thing it doesn't include is a letter or a visit to a store.

**CALL ABANDONS** Also called **ABANDONED CALLS**. Call Abandons are calls that are dropped by the calling party before their intended transaction is completed.

The call may be dropped at various points in the process (i.e., while on hold, while dialing, etc.). The point in the call at which the call is abandoned will have varying impacts on a telephone system. Many callers will hang up as soon as they realize they've reached an automated system and not a person. For systems that expended significant energy in setting up to answer a call, a large percentage of call abandons can negatively impact the call capacity of the system.

**CALL ACCOUNTING SYSTEM** An automated system for recording information about telephone calls, organizing that information and upon being asked, preparing reports — printed or to disk. The physical system itself consists of a computer, a data storage medium, software and some mechanical method of attaching itself to a telephone system.

The information which it records (or "captures") about telephone calls typically includes the extension from which the call is coming, which number it is calling (local or long distance), which circuit is used for the call (AT&T, Sprint, a tie line to another office, etc.), when the call started, how long it lasted and for what purpose the call was made (which client? which project?).

A call accounting system might also include information on incoming calls — which trunk was used, where the call came from (if ANI or interactive voice response was used), which extension took the call, if it was transferred and to where and how long it took.

There are nine basic uses for call accounting systems in the call center:

1. Controlling Telephone Abuse. It's the 90-10 rule. 10% of your people sit on long distance calls all day to their friends and family. The others work. Knowing who's calling where and how much they're spending is useful.

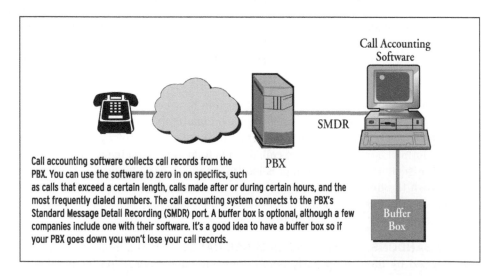

Call accounting software collects call records from the PBX. You can use the software to zero in on specifics, such as calls that exceed a certain length, calls made after or during certain hours, and the most frequently dialed numbers. The call accounting system connects to the PBX's Standard Message Detail Recording (SMDR) port. A buffer box is optional, although a few companies include one with their software. It's a good idea to have a buffer box so if your PBX goes down you won't lose your call records.

2. Controlling Telephone Misuse. A particular five minute call between two major cities could cost either five cents a minute, or $1 a minute, depending on how it was dialed and what line it went out over. That's a 20-fold difference! Sometimes the phone system makes the dialing decision. Sometimes the person makes the dialing decision. Whoever's doing it can be wrong. A call accounting system is a good check to see if you're spending money needlessly.

3. Allocating telephone calling costs among departments and divisions. Telecommunications — including sending voice, data, video and imaging — are some of your biggest expenses. (In a call center phone costs are the second largest ongoing expense, after labor.) They're a cost that should be allocated to the products you're making, or the departments or divisions in your company. Telephone costs can determine which product is profitable, and which isn't. Item: A software company recently dropped one of its three "big" software packages because phone calls for support got too expensive.

4. Sharing and resale of long distance and local phone calls, as in a hotel/motel, hospital, shared condominium, etc. Someone's got to send out the bills. And it's not the phone company. In fact, with a call accounting system you can be your own phone company!

5. Personnel evaluation and motivation. Which employees are doing better at being productive on the phone (however you define "productive"). You want them to get on and off the phone fast? Or you want them to stay on and coddle your customers? You can now correlate phone calls with income — from service or just straight-out sales. If you don't use an ACD or call center management software, a call accounting system can provide this information.

6. Network optimization. Two fancy words for figuring which is the best combination of MCI Worldcom, AT&T, Sprint (etc.) lines. And which is the best combination of all the various services each offer. A rule of thumb: There's a 20-fold difference in per minute telephone calling costs between any two major cities in the US. And — amazingly — you won't hear any difference in quality, despite the huge difference in price.

7. Phone system diagnostics. Is the phone system working as well as it should? Are all the lines working? Are all the circuit packs (circuit cards) working? Call accounting systems can tell you which lines you're getting no traffic on. Or which line carried the 48 hour call to Germany (it's happened). Either way, you can figure quickly which lines are working and which aren't.

8. Long distance bill verification. Was the bill we received from our carrier accurate? Often it isn't. In fact, there's no such thing as an accurate phone bill. That's an oxymoron. Using your call accounting systems to check your long distance gives you some peace of mind.

9. Tracing Calls. True story: Every third or fourth Friday afternoon a large factory in the South received bomb threats. They'd clear the factory, search the factory and not find anything. By the time they'd checked, it was too late to start up production. One day they checked their call accounting records. The calls were coming from a phone on the factory floor. The whole thing was a ruse to get an afternoon off. And now that many phones give you the number of who's calling, call accounting systems are turning out to be great for checking the effectiveness of regional ad campaigns, figuring the profitability of direct mailings and even figuring the profitability of individual customers.

**CALL ATTEMPT** A try at making a telephone call to someone. Tally up your outbound call attempts and compare them to completions and you'll have some idea of corporate frustration and thus, the need for more lines, more phone equipment or a predictive dialer. Do the same for inbound call attempts and completions, and you'll have some idea of the frustration your customers have trying to reach you.

**CALL BLENDING** This refers to the process of combining inbound and outbound call handling capacity to balance the load of each. It used to be that inbound and outbound were separate; each kind of call was handled by its own group. Outbound calls often went through a dialing processor separated from the main inbound switch. But outbound calls often result in calls coming back later (especially in business-to-business sales, or collections applications).

Call blending allows the same person (or group) to handle calls that go in both directions. Why would you want to do that? Because the volume of calls varies greatly depending on a number of factors, including time of day, day of week, business conditions, and so on. The idea is to keep the calls at a constant level.

Call blending automatically transfers staff members between outbound and inbound programs as call volumes change. Some predictive dialers let you choose which workstations will be used for call blending, to avoid training of every staff member.

**CALL-BY-CALL ROUTING** A method of "pre-routing" calls bound for one of a several call centers on a linked network. In a call-by-call environment, each call is routed individually by a routing processor at the customer's site, analyzing both the instant call center data as well as information provided by the network to determine how that call should be handled. Each call is examined as a separate processing event — which distinguishes it from routing based on percentage allocation, where calls are considered more as batches.

In call-by-call, every call is a discrete occurrence for processing, with its own qualities and its own query against network and call center status. It allows for quick reactions to changing conditions, and makes for a powerful tool in situations where call volume and staffing fluctuate a great deal. It also let you see detailed composite reporting of how calls were handled across centers, including very specific data about which calls were sent where, and what their dispositions were.

Call-by-call does have disadvantages, however. When one or more call centers in the network are fronted by an IVR system, it is the IVR which receives the call, instead of the targeted agent or group. The caller doesn't make a self-routing decision until after the call has been delivered. This removes the pre-call routing discretion that the distributed center network was relying on for efficiency.

**CALL CENTER** A place where calls are placed, or received, in high volume for the purpose of sales, marketing, customer service, telemarketing, technical support or other specialized business activity. One early definition described a call center as a place of doing business by phone that combined a centralized database with an automatic call distribution system. That's close, but it's more than that:

• Huge telemarketing centers

• Fundraising and collections organizations

• Help desks, both internal and external

• Outsourcers (better known as service bureaus) that use their large capacity to serve lots of companies

• Order departments, like the ones at Lands' End, L.L. Bean and other catalogs.

Estimates of the number of call centers in North America range from 20,000 to as high as 200,000. The reality is probably somewhere around 80,000 to 100,000 depending on what you consider a call center. Some experts believe that you shouldn't count centers below a certain number of agents (or "seats"). We believe in the widest possible definition, all the way down to micro-centers of four or five people. Why? Because those centers face many of the same kinds of problems on a daily basis as their larger cousins: problems of training, staffing, call handling, technology assessment, and so on. Those smaller centers have to put the same kind of face forward to the customer as the largest centers, in order to remain competitive. And more often than not, those small center become medium-sized centers over time.

Call centers are generally set up as large rooms, with workstations that include a computer, a telephone set (or headset) hooked into a large switch and one or more supervisor stations. It may stand by itself, or be linked with other centers. It may also be linked to a corporate data network, including mainframes, microcomputers and LANs.

Call centers were first recognized as such in their largest incarnations: airline reservation centers, catalog ordering companies, problem solvers like the GE Answer Center or WordPerfect's customer support services. Until the early 1990s, only the largest centers could afford the investment in technology that allowed them to handle huge volumes (the ACD). More recently, with the development of PC LANs, client/server software systems, and open phone systems, any call center can have

**39**

an advanced call handling and customer management system, even down to ten agents or less.

As companies have learned that service is the key to attracting and maintaining customers (and hence, revenue), the common perception of the call center has changed. It is rarely seen as a luxury anymore. In fact, it is often regarded as a competitive weapon. In some industries (catalog retailing, financial services, hospitality) a call center is the difference between being in business and not being in business. In other industries (cable television, utilities) call centers have been the centerpiece of corporate attempts to quickly overhaul service and improve their image.

We believe that any company that sells any product has a call center, or will shortly have one, because it is the most effective way to reach (and be reached by) customers.

**CALL CENTRE** The spelling of "call center" in most English-speaking countries outside the United States.

**CALL COMPLETION RATE** The ratio of successfully completed calls to the total number of attempted calls. This ratio is typically expressed as either a percentage or a decimal fraction.

**CALL CONTROL** Call control is the term used by the telephone industry to describe the setting up, monitoring, and tearing down of telephone calls. There are two ways of doing call control. A person or a computer can do it via the desktop telephone or a computer attached to that telephone, or the computer attached to the desktop phone line (i.e. without the actual phone being there). That's called First Party Call Control. Third-party call control controls the call through a connection directly to the switch (PBX). Generally third-party call control also refers to the control of other functions that relate to the switch at large, such as ACD queuing.

**CALL CONTROL PROCEDURE** Group of interactive signals required to establish, maintain and release a communication.

**CALL CONTROL SIGNAL** Any one of the entire set of interactive signals necessary to establish, maintain, and release a call.

**CALL DATA** Call data refers to any information about a phone call that is passed by a switch to an attached computer system. Call data is usually used by a computer telephony application to process the call more intelligently. Call data may be passed In-Band (over the same physical or logical link as the call, usually through tones), or Out-Of-Band (over a separate link, usually a serial link).

Call data may also be passed as part of the telephone network control links, such as SS7 (Signaling System 7) links. In addition to information about the call, its status and even control over the call can be available as part of the call data link services.

Call data almost always includes what number dialed the call (ANI) and/or what number was called (DNIS). More complex call data links used for "PBX integration" may also indicate why the call was presented (such as forwarded-on-busy), or tell what trunk the call is coming in on.

Full blown computer telephony links, such as are now being offered by many switching vendors, enhance the call data path, providing additional status information about calls and can even provide a level of call control to the attached computer telephony system.

**CALL DETAIL RECORDING (CDR)** A feature of a telephone system which allows it to collect and record information on outgoing and incoming phone calls — who made/received them, where they went, where they came from, what time of day they happened, how long they took, etc. Sometimes the data is collected by the phone system; sometimes it is pumped out of the phone system as the calls are made. Whichever way, the information must be recorded elsewhere — dumped right into a printer or into a PC with call accounting software. See also CALL ACCOUNTING SYSTEM.

**CALL FLOW** A term that has the same general meaning as call routing, but with some different shades of meaning. Call routing, especially as defined by a routing table, generally refers to routing within a single ACD or telephone switch. It tells which agent groups or gates a call goes to under certain circumstances. Call flow is the path the call takes from the time it enters the call center until the time it leaves. It may be answered by an ACD, then transferred to an IVR system, then sent back to the ACD and off to a remote call center where it is answered by the agent with the third-best skills match. All of the decisions that went into sending that call to all those different places make up the call flow.

**CALL-ME BUTTONS** These days, call centers are seeing calls come in from a new kind of front-end device: the web site. Just as IVR has delivered calls to particular queues and groups based on caller-entered information, so can a web site, with its many options for collecting customer data.

In the most common technical implementation of call center/web site combination, a clickable button is placed on a company's web site that invites the web surfer to click through "to speak to a representative." Sometimes the button leads to an interactive form, where the customer is instructed to fill out information about what he wants, and include a phone number and best time to receive a call. When the form is posted, a request is sent to the queue for an outbound call back to the customer.

Or, it could initiate a call using internet telephony, hooking a call center rep up with the customer, speaking through her computer, without dropping the customer's internet connection. This doesn't happen often, though, due to the paucity of IP telephony-enabled consumer computers.

It's important to note that call-me buttons are a solution in search of a problem — few companies have found a good way to implement them that benefits both the customer and the call center.

**CALL MIX** Call mix is the pattern of call types (each call type defines what the caller does for that call) that goes into creating a busy hour call profile or other call profile. A voice mail system's busy hour call profile call mix may look like this:

- 10% call abandons,

- 20% login and send one message,

- 30% login and listen to one message, and so on.

n a call center "call mix" also means the particular blend of call types or functions received by a certain center or at a certain time. For example, 30% customer service, 60% orders and 10% technical support.

**CALL PRIORITY** A term for the value assigned to an incoming call, from highest priority to lowest.

**CALL PROCESSING** The movement of a call to its intended point, through all its automated twists and turns. Call processing is the set of instructions that are used to deal with a call, and the act of dealing with the call. You more often hear of a call processing system (read: switch, plus its software) than the "process" itself. The act of call processing is roughly equivalent to switching, except that call processing implies that the connection is switched according to those pre-set instructions.

**CALL PROCESSING UNITS** The systems that scan the trunk and/or station port for incoming calls and set up the connection when it occurs.

**CALL PROGRESS** The status of the telephone line; ringing, busy ring/no answer, voice mail answering, telephone company intercept, etc. See CALL PROGRESS ANALYSIS and CALL PROGRESS TONE.

**CALL PROGRESS ANALYSIS** Once a call is dialed, several things can happen to it: The phone rings on the other end. An answering machine answers. A fax machine may answer. There might be a busy or operator intercept. Call progress analysis is the process of figuring out which is occurring as the call progresses. This analysis is critical if you're trying to build an automated system, like an interactive voice response system.

**CALL PROGRESS TONE** The sounds generated by the telephone network that signal what is happening with the call. For example, there is a tone that tells you the phone on the other end is ringing. Another tone tells you the phone on the other end is busy, still another tone (these days more likely a message) that tells you the telephone network itself is busy. When the number you have dialed has been changed or is out of service, a series of three tones is heard before the recorded message.

When you use a predictive or other automated dialing system, these tones suddenly have increased importance. The ability of your dialing system to recognize these tones and respond accordingly, without involving an agent, greatly increases the efficiency of the system.

**CALL RECORD** Information about a call (extension or position, length, time-of-day, number dialed) recorded by a PBX or ACD. These records are the basis of call center management software and telecommunications management software systems.

**CALL REPORTING** In an outbound environment, detailed accounting of telemarketing activity measured by agent, group of agents, campaign, region, or other key factor. Similar reports are also important for keeping tabs on inbound marketing. Good reporting capabilities are a critical feature for both dialers and ACDs.

**CALL ROUTING** Literally, the list of choices a user sets up within an ACD for where to send the incoming calls. You might define a path for calls that last a certain amount of time, or define options — if a caller presses "0", send the call to an operator, if she presses "1" send her to technical support. The routing table will reflect different campaigns handled in your call center, enabling the ACD to send calls to the right agent groups or departments. Within the ACD, you can use more sophisticated criteria to direct the call, like skills. For example, calls that originate from a particular country are sent to agents with certain language abilities. Parsing incoming ANI or DNIS data for routing is increasingly important.

**CALL SECOND** A unit for measuring communications traffic. Defined as one user making one second of a phone call. One hundred call seconds are called "ccs," as in Centum call seconds. "CCS" is the U.S. standard of telephone traffic. 3600 call-seconds = 1 call hour. 3600 call-seconds per hour = 36 CCS per hour = 1 call-hour = 1 erlang = 1 traffic unit. See also ERLANG and TRAFFIC ENGINEERING.

**CALL SEQUENCER** A call sequencer, also called an Automatic Call Sequencer, is a piece of equipment which attaches to a key system or a PBX. The call sequencer's main function is to direct incoming calls to the next available person. It typically does this by causing lights on telephones to flash at different rates. The light with the fastest flashing is the one whose call has been waiting longest. This call is answered first. Call sequencers also might answer the phone, deliver a message and put the person on hold. They might keep statistical tabs of incoming calls, how fast they were answered, how long the people waited, how many people abandoned (hung up while they were on hold waiting for their call to be answered by a human being).

Call sequencers are usually simple and inexpensive. Better, but much more expensive devices for answering incoming phone calls are Automatic Call Distributors. See AUTOMATIC CALL DISTRIBUTOR, UNIFORM CALL DISTRIBUTOR.

**43**

**CALL SETUP TIME** The amount of time it takes for a circuit-switched call to be established between two people or two data devices. Call set-up includes dialing, wait time and time to move through central offices and long distance services. You don't pay for call set-up, but you will need extra lines to take care of it.

**CALL SHEDDING** In many states, laws require that a real live person be available for a phone call being outdialed with an automated device. The reason these laws were enacted is because a lot of times call center managers had their predictive dialers going so crazy in search of human answerers that they didn't have an agent ready when someone did actually pick up the phone. Most systems simply hang up the connection when this occurs. This is called "call shedding."

**CALL TABLE** A term that refers to the list of records defined by a downloaded pre-processing script. Another way of describing the call list being sent to the predictive dialer.

**CALL TYPE** A term used in Rockwell ACDs. A portion of your call center traffic corresponding to one or more ACD gates or splits. This division of the total ACD traffic is the level at which forecasting and scheduling are done. At setup time, each Call Type is defined in the ACD by a unique three-letter code and specific gate or split number(s) that identifies the corresponding ACD report data.

**CALL VECTOR** See VECTOR.

**CALL VOLUME** 1. The number of calls that can be handled by an ACD in a given period, or 2. The number of calls that come into a call center in a given period.

**CALLCENTER** A popular ACD system once made by Aspect Communications.

**CALLED PARTY PAID SERVICE** Another way of saying "toll-free service." Using this term emphasizes that, in fact, the calls are not free, but are charged to the party who receives them. Called party paid service numbers begin with the code 800, 888 or 877. In the future, they may also start with 866, 855, 844, 833 or 822. See TOLL FREE SERVICE.

**CALLER ID** On an incoming phone call, the information you get that lets you know where the call is coming from before you answer it — that is, the telephone number of the person calling you. Caller ID is the consumer "brand name" for the local phone company variety of this service. It's really called CLASS (short for Custom Local Area Signaling Services). In CLASS, the calling party's phone number is passed to your phone between the first and second ring signaling an incoming call. The simplest Caller ID devices show you the originating phone number of the incoming call. If you have the appropriate software, though, you can match that phone number with a caller's name (and other information straight from your database). That gives Caller ID much more power as a call center tool.

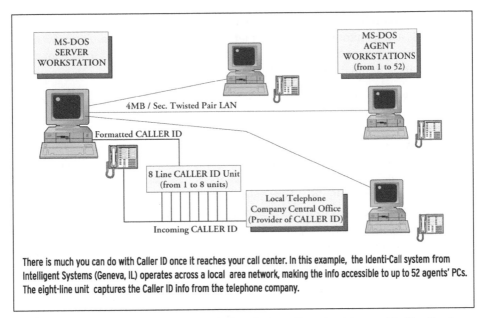

```
MS-DOS
SERVER
WORKSTATION
```

```
MS-DOS
AGENT
WORKSTATIONS
(from 1 to 52)
```

4MB / Sec. Twisted Pair LAN

Formatted CALLER ID

```
8 Line CALLER ID Unit
(from 1 to 8 units)
```

```
Local Telephone
Company Central Office
(Provider of CALLER ID)
```

Incoming CALLER ID

There is much you can do with Caller ID once it reaches your call center. In this example, the Identi-Call system from Intelligent Systems (Geneva, IL) operates across a local area network, making the info accessible to up to 52 agents' PCs. The eight-line unit captures the Caller ID info from the telephone company.

In fact, the more important variety of Caller ID (to call centers) is provided by the long distance carriers in the form of ANI — automatic number identification, chiefly on 800 calls. It delivers the calling party's telephone number, which can then be looked up in your company's database. This is what enables repeat customers to be greeted with "Hello, Ms. Smith, how do you like the shirt you bought from us last week?"

**CALLER INDEPENDENT VOICE RECOGNITION** Having a voice response unit recognize the voice of a caller without having been trained on the caller's voice. Most commonly referred to as SPEAKER-INDEPENDENT VOICE RECOGNITION.

**CALLED LINE IDENTIFICATION SIGNAL** A sequence of characters transmitted to the calling terminal to permit identification of the called line.

**CALLING LINE ID** Another term for Caller ID, especially on local calls. When referring to call information from a toll-free service, the more appropriate term is ANI (AUTOMATIC NUMBER IDENTIFICATION).

**CALLING NUMBER DISPLAY** The screen on your phone (or attached peripheral device) that shows you ANI or Caller ID data on incoming phone calls — the phone number, or the name of who's calling.

**CALLING PARTY'S NUMBER** See CPN.

**CALLPATH** Software from IBM which let you integrate one of IBM's computer systems with selected telephone systems. For example, CallPath/400 allowed integration between IBM's AS/400 computer system and a variety of telephone switches.

**45**

The integration lets the computer application redirect inbound calls, initiate outbound calls and monitor call progress. It lets the system perform simultaneous computer screen and voice transfers for ANI or other applications. Used to be a lot more important back in the days of closed phone systems.

**CALLPOWER** A Rockwell ACD term. An integrated voice and data workstation for use in combining ACD capabilities with host computer database management.

**CAMPAIGN** A project or program running in your call center. It generally refers to an outbound project, though not exclusively. The most common use of campaign is in outbound dialing, where automated dialers and software programs often tout their ability to "run multiple simultaneous campaigns." That means that you can have some agents calling to sell cereal, and others calling to sell toothpaste, at the same time, on the same equipment. It is a useful feature, especially for service bureaus and other outsourcers.

**CAMPAIGN FLOW** This is a set of CALL TABLES linked together to provide a continuous outbound calling list.

**CAPACITY** The information carrying ability of a telecommunications facility. What the "facility" is determines the measurement. You might measure a data line's capacity in bits per second. You might measure a switch's capacity in the maximum number of calls it can switch in one hour, or the maximum number of calls it can keep in conversation simultaneously. You might measure a coaxial cable's capacity in bandwidth.

**CARBON-COMPATIBLE** A headset that is compatible with a telephone that uses a carbon microphone as its receiver. A carbon-compatible headset can usually run off of the line current from the telephone. Other types of headsets (and telephones) need an amplifier and an external power supply.

**CARD (STATION, TRUNK)** Most modern telephone systems consist of cabling, telephones and a cabinet that contains some basic electronics and lots of slots. In these slots go station (telephone) and trunk (telephone company line) cards. Some systems have universal slots so you can create your own balance of stations to trunks. Other systems have dedicated slots for each type of card. These kinds of cards are printed circuit boards, just like the add-in boards for PCs.

When you buy a telephone system there should be some empty slots so you can add cards as your company grows. Once all the slots are filled up you have to buy another system or add an expansion cabinet, depending on the system.

**CARRIER** A company which provides communications circuits. Carriers are split into "private" and "common." A private carrier can refuse you service. A "common" carrier can't. Most of the carriers in our industry — your local phone company, AT&T, MCI Worldcom, Sprint, etc. — are common carriers. Common carriers are regulated. Private carriers are not.

46

**CARRIER BYPASS** A link between your center and the long distance phone company's switching office that doesn't have the local phone company sitting in the middle. A bypass is done to save the customer or the long distance company money. Bypass is also done to get service faster. Sometimes the local phone company simply can't deliver fast enough.

**CASE-BASED REASONING** A method of solving problems that uses prior experience to find the most appropriate solution to a given problem. It's one of several methods used in help desks and for technical support. A software system that includes it records details of problems (each called a "case") and their solutions. The object of it is to keep you from having to solve the same (or similar) problems time and again. The longer you use such a system, the better it becomes at finding answers, because its "case base" is larger and more varied.

**CATI** Computer Assisted Telephone Interviewing, a market research term for a specialized outbound application, and the class of software that makes it possible. It's extremely similar to telemarketing, in that you call from a list and use a script to guide you through the call and record answers.

**CCITT** Comite Consultatif Internationale de Telegraphique et Telephonique. Yes, that's French. An international telecommunications standards-making organization based in Geneva. Keep an eye out for these specs if you plan to use a product or service internationally, especially in Europe.

**CCMS** A Siemens term. It stands for Call Center Management System. It also sometimes (but rarely) stands for generic Call Center Management Software.

**CCS** 1. One hundred call seconds or one hundred seconds of telephone conversation. One hour of telephone traffic is equal to 36 CCS (60 x 60 = 3600 divided by 100 = 36) which is equal to one erlang. CCS are used in network optimization. Lee Goeller calls CCS an obsolete traffic unit. He says "When given traffic in CCS, always divide by 36 immediately. It is not obvious that 5 trunks cannot carry 185 CCS, but you don't have to be a rocket scientist to know that you can't average 5.14 calls on a five trunk group." That's good advice. 2. Cumulative call seconds — a measure of trunk occupancy.

**CCS/SS7** A Bellcore term for Common Channel Signal/System Signaling 7. See SIGNALLING SYSTEM 7.

**CDR** Call Detail Recording (as in call accounting) or Call Detail Record, as a record generated by customer traffic later used to bill the customer for service. See CALL ACCOUNTING.

**CDR EXCLUDE TABLE** A table listing local central office codes which are deliberately ignored by a call accounting system.

**CENTRAL OFFICE** Telephone company facility where subscribers' lines are joined to switching equipment for connecting other subscribers to each other, locally and long distance. Also called CO, as in See-Oh. Sometimes the term central office is the same as the overseas term "public exchange."

**CENTREX** All telephone switches work on the same general principle. They take calls coming in and route them to the proper extension. At home, your local telephone company switch (central office) routes calls coming into it to the proper phone numbers. Same principle.

With Centrex you use part of your local telephone company (central office) switch as your office telephone system. Incoming calls can work just like residential service: someone calls your phone number and gets your office directly. But you also get hold, transfer and other features. Someone can call your company's general number and the switchboard attendant can switch the call to you.

There are many advantages, but most boil down to the fact that the phone system is the phone company's problem. They have it on their premises and taking care of it is their problem. The disadvantage is you have to go through your local telephone company to make any changes — instead of working on the fly as many companies can with their PBXs.

However: Just try building a call center using Centrex. See how far you get. The single biggest reason for buying a PBX over Centrex (in this context) is that you can graft ACD features onto a PBX.

Some local telephone companies offer Centrex ACDs (under many names). The advantages and disadvantages are the same as for regular Centrex. Besides the expense (this is a rent, not buy arrangement) the biggest drawback, is, of course, that with a Centrex ACD your local phone company has total control over your center's switching and routing. A rare few call center managers are comforted by this. Most are horrified.

The biggest interest in Centrex ACDs these days is not for the feature itself, but for your ability to team it with on-premises technology (mostly software based) and create a sophisticated, flexible ACD system.

**CENTREX CALL MANAGEMENT** A Centrex feature that provides detailed cost and usage information on toll calls from each Centrex extension, so you can better manage your telephone expenses.

**CENTUM CALL SECOND** 1/36th of an erlang. The formula for a centum call second is the number of calls per hour multiplied by their average duration in seconds, all divided by 100.

**CGI** Common Gateway Interface. A World Wide Web gateway.

**CHARGES AGAINST REFUNDS** The amount set aside from the proceeds of a 900 service to cover "uncollectables" and make the phone company whole. It ranges from as little as 1% of the budget for business-oriented information programs, up to 50% for adult and talk lines.

Customers who refuse to pay can't be disconnected for non-payment of 900 bills. In many cases, the phone company will remove the charge from the customer's bill and pass it on to the information provider. (The phone company will charge the IP for billing and transporting the call whether or not the customer pays his end.)

**CHASE** A conduit for electrical, phone or computer cable. Many types of call center furniture have this conduit built right in to the unit. Make sure the conduit is large enough for your needs (fiber optics and some types of computer cable don't bend well) and that the chase is easy to get to if you need to get at the cable.

**CLASS** Custom Local Area Signaling Services. Class consists of number-translation services, available within a local exchange. CLASS is a service mark of Bellcore. Caller ID is the most important service available as part of CLASS. See CALLER ID.

**CLEAN** Correcting data on a list or database and removing undeliverable addresses or incorrect telephone numbers. A clean list is one with no duplication of records, with every address correct and telephone number reachable, and where every entry on the list meets the list criteria (if it's a list of millionaires, they are all millionaires).

**CLICK-TO-CALL** A button placed on a web page that initiates a call into a call center, usually by dropping the customer's internet connection and connecting a voice call. See CALL-ME BUTTONS.

**CLID** Caller Line IDentification. Once available only from your local telephone company, today CLID is the standard for providing calling party information over telephone lines. CLID must be ordered from your telecommunications carrier and almost always costs extra. You need a device to receive CLID and display it. That device can be the little box you buy at Radio Shack or your properly-equipped ACD. Programming must be done to have the phone number initiate a database look-up giving you whatever info you have stored, keyed to the phone number. The telephone carrier provides only the telephone number, and perhaps a name. You must do the rest.

**CLIENT** Clients are devices and software that request information. Clients are objects that uses the resources of another object. A client is a fancy name for a PC on a local area network — it is the "client" of the server. (Clients are also humans who buy things from your company, but you knew that, didn't you?)

**CLIENT APPLICATION** Any computer program making use of the processing resources of another program.

**CLIENT/SERVER** Client/server computing refers to a system that splits the workload between desktop PCs (called "workstations") and one or more larger computers (called "servers") joined on a local area network (LAN). The advantage of the client/server model has over the model it replaced (mainframes and dumb terminals) is that it speeds up access to the pieces of information people need most.

In a call center, the clients are agent desktops. The servers are simply the application servers. In an order entry application, for example, the client can spend the processing power on an easy-to-use graphical interface, because most of the database processing is done at the server. It's faster, because <I>some of the processing is done locally, when necessary. It keeps large bottlenecks from slowing down everybody on the system.

The server can be a minicomputer, workstation, or microcomputer with attached storage devices. A client can be served by multiple servers.

**CLOSED ARCHITECTURE** Proprietary design that is compatible only with hardware and software from a single vendor of single product family. Contrast with OPEN ARCHITECTURE.

**CLOUD LEVEL** At the level of the public switched telephone network. Or, sometimes, a service provided by your long distance carrier. For example, "cloud level call allocation" would be a divvying-up of calls performed in the public switched telephone network, probably by your long distance carrier. The term comes from telecommunications diagrams where the public switched telephone network is represented by a big cloud. Also see PSTN. The term is sometimes used to indicate a long-range or high level point of view. "Her view of call center operations is cloud level." It has connotations of "head in the clouds," theoretical and not quite down to earth.

**CLUSTER** A cluster is a particular style of call center work area, where a number of desks are "clustered" around a central core in a star pattern. (They are not "cubicles" because they are more triangular than square.)

**CMS** 1. Call Management System. This is the Lucent label for their inbound call distribution management reporting package. 2. Call Management Services. Canadian term for local calling features based on CLID (Calling Line Identification).

**CMT** Conflict Management Training. Not, as you might imagine, a seminar on how to direct a war or settle a strike. It usually refers to something much lower key. It is a type of training given to customer service representatives, especially those that frequently handle complaints. The conflict referred to here is between the customer and the company or perhaps between the customer's needs and the company's needs. The "management" is the techniques and skills the rep brings to resolving those conflicts.

**CNG** Calling Tone. The sound produced by virtually all Group 3 fax machines when they dial another fax machine. If your agents recognize the sound, they may be able to transfer misplaced calls to a fax machine. At the very least, they will no longer fear phone calls from alien spaceships or hostile computers.

**CO** See CENTRAL OFFICE.

**CO ACD** See CENTREX.

**COAXIAL** A well-insulated and protected wire that is used to carry high-speed data (such as between a host computer and terminal). Your cable TV service comes into your house on coaxial cable.

**COLD CALL** The first contact with a potential customer in the sales process. The person has shown no interest in your product or service. You have no good reason to believe they might be interested in what you have to offer. But you call them, on the telephone or in person, to see if there is an interest. Cold calling promises lots of rejection and is generally dreaded by both telephone and field salespeople.

**COM (COMPONENT OBJECT MODEL)** Microsoft's framework for developing and supporting "component objects" — modular bits of code that can be used for fast creation of add-on applications in large software systems. For example, you can have a call tracking and routing system that consists of many COM modules — and swap in or out modules for particular vertical industries like banks depending on customer needs. The COM model is supposedly a faster one for upgrading complex software systems. See DCOM.

**COMMON CARRIER** A company that furnishes communications services to the general public. It is typically licensed by a state or federal government agency. A common carrier cannot refuse to carry you, your information or your freight as long as you conform to the rules and regulations as filed with the state or federal authorities. See OTHER COMMON CARRIER.

**COMMON CARRIER BUREAU** A department of the Federal Communications Commission responsible for recommending and implementing regulatory policies on interstate and international common carrier (voice, video, data) activities.

**COMMUNICATIONS SERVER** A computer platform used in a telecommunications network much as a data server is used in a LAN (local area network). The processing hub of a communications network, especially your company's internal network if you use an UnPBX or server-based ACD. See UNPBX and SERVER-BASED ACD.

**COMMUNICATOR** A British term. An alternative, and probably more meaningful, name for a telebusiness agent, or as he or she would be called in North America — an agent.

**COMPILED LIST** A list created from entries in telephone books or "yellow pages," industry directories or public records. When you are looking for completeness or saturation in a list and don't care that much about evidence of previous direct mar-

**51**

keting responsiveness, look to compiled lists. Compiled lists strive to give you every listing in a chosen category, for example, every household in Woodbridge, NJ or every computer programmer in Silicon Valley.

**COMPRESSION (VOICE/SPEECH)** In order to store voice in a smaller "space" on a computer disk or tape or to transmit it faster over telephone lines, it is reduced by one or several methods. These methods include taking out the pauses and making a "sketch" of the voice, reducing it to its most vital elements. The signal must be expanded for it to be comprehensible.

**COMPUTER** This is a definition straight from AT&T Bell Laboratories. "An electronic device that accepts and processes information mathematically according to previous instructions. It provides the result of this processing via visual displays, printed summaries or in an audible form."

**COMPUTER-AIDED DIALING** A newer (and allegedly less offensive) term for predictive dialing. See also PREDICTIVE DIALING.

**COMPUTER-BASED LOOKUP** A telephone number matching service that enhances lists using computers to run through databases and attach phone numbers to names and addresses. It is fast and relatively inexpensive, but computers tend to miss some matches that people would find obvious.

Companies that offer computer-based lookup services offer a number of different services that play to their strengths, particularly with larger lists, like adding demographic information.

**COMPUTER TELEPHONY** A term that describes the process of applying computer intelligence to telecommunications devices, especially switches and phones. The term covers many technologies, including computer-telephone integration through the local area network, interactive voice processing, voice mail, automated attendant, voice recognition, text-to-speech, fax, simultaneous voice data, signal processing, videoconferencing, predictive dialing, audiotext, collaborative computing and traditional telephone call switching.

Or, you can look at it from the other side, and say that it's the process of incorporating telecom devices into computer systems. Which view you take probably depends on what industry you came from.

**CONCATENATED SPEECH** A method of creating a voice message from smaller digitized pieces (such as words, phrases or parts of words). This is the type of message you get when you call an automated system for your bank balance or you call directory assistance for a telephone number. Concatenated speech is best when you know you'll be dealing with a small set of stock words or phrases.

**CONDITIONAL ROUTING** The routing of calls, usually among or between call

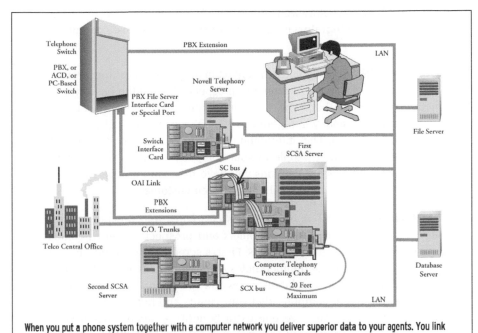

When you put a phone system together with a computer network you deliver superior data to your agents. You link them to huge data resources, and you make it easier to add peripherals like voice systems or fax servers.

centers, based on criteria that you might decide to change. That is, one way to route calls is to send 60% of calls to Los Angeles and 40% to New York, all the time. But conditional routing would say that if New York is experiencing particularly heavy traffic, then the system should send New York-bound calls to Los Angeles instead, until traffic flattens out. There are many criteria you can use to set conditions for routing, including time of day, type of call, traffic patterns, availability of agents with particular skills, and so on. Less often, you could see this term used to describe calls routed within a particular center, among agent groups or even individual agents.

**CONFLICT MANAGEMENT TRAINING** See CMT.

**CONSOLE** A specialized telephone. Telephones (or station sets) used by sales or customer service representatives behind an ACD are often called consoles, as are the specialized sets used by PBX attendants. See ATTENDANT CONSOLE.

**CONSULTIVE SELLING** This selling technique emphasizes the needs and wants of the customer. The seller asks the prospect a series of questions to find out if the seller's product or service will help the prospect. The questions are probing and open-ended. Consultive selling involves listening more than talking; objections are addressed, not overcome. The sales pitch focuses on how the sellers product can meet the *specific desires* (or allay the particular fears) of the prospect.

**CONTACT GATEWAY** A Rockwell ACD term. A hardware/software package that acts as a gateway to integrate various ACD, voice response, host computer and live agent processes.

**CONTACT HISTORY** When you use a contact management program, you are accumulating information about a customer or prospect. Beyond the basic identification information (phone number, type of business, etc.) you are also collecting data about what you have said, what literature you have sent, what interests the person has expressed, and so on. That information is the contact history. It is of vital importance, because it lets other people in your company speak to that customer with complete knowledge of what dealings you have had. And if a salesperson leaves, the company doesn't lose all the valuable data on customers.

**CONTACT MANAGEMENT** A business has customers and prospects. Software to "manage" your interactions with customers and prospects is called contact management software. It has three elements: First, a screen or two of information about that contact (address, phone number, notes about your conversations). Second, the ability to print lists, and letters, labels, faxes and so on. And third, a tie-in with your phone system to let your computer dial your clients and fax them stuff. With many newer phone systems, you have one extra benefit — namely when your phone rings, your contact management software will receive the calling phone number and pop up a screen or two about your contact. This way you'll be a little prepared before you answer the phone.

**CONTENT PROCESSING** Voice processing is the broad term made up of two narrower terms — call processing and content processing. Call processing consists of physically moving the call around. Think of call processing as switching. Content consists of actually doing something to the call's content, like digitizing it and storing it on a hard disk, or editing it, or recognizing it (voice recognition) for some purpose (like using it as input into a computer program).

**CONTINUOUS SPEECH** A type of voice recognition which allows users to speak naturally, that is, to run words together. This is much more sophisticated than DISCRETE SPEECH and requires a lot more horsepower from the computer it runs on.

**CONTOUR HEADSET** Also known as an "around-the-ear" headset. A contour headset gets its stability from a piece that wraps around the back of your ear rather than with a band over your head.

**CONVERSATION TIME** A term used in automated, outbound dialing. It is the time spent on a conversation from the time the person at the other end picks up to the time you or he hangs up. Conversation time plus dialing, searching and ringing time equal the time your circuit will be used during an automated, outbound call.

**CORRELATION** In statistics and market research, a proved link between two or more data characteristics. For example, interest in the game of golf might correlate with household income. A correlation can be negative — people who have cats might have low interest in dog food.

**COST-PER-CALL ANALYSIS** A measure of the profitability of a call center, often expressed in reports generated by call center management software packages. Cost-per-call takes into account the cost of labor, phone service and equipment reflected against revenue generated. This measure is used more and more as an alternative to "performance measurement" that accounts more strictly for the length of time spent on calls.

**COST PER INQUIRY** Often abbreviated CPI.

**COST PER ORDER** Often abbreviated CPO.

**COST PER THOUSAND** Often abbreviated CPM.

**COUPONING** The practice of advertising a 900 number that customers call to obtain special coupons for products, usually consumer goods. It has three goals: 1) Capturing data on pre-qualified customers that can be used for market research and list-building; 2) Tracking buying habits by matching the coupons to the callers; and 3) Building customer loyalty while encouraging repeat purchases.

**CPE** Customer Provided Equipment, or Customer Premise Equipment. Originally it referred to equipment on the customer's premises which had been bought from a vendor who was not the local phone company. Now it simply refers to telephone equipment which resides on the customer's premises. "Premises" might be anything from an office to a factory to a home.

**CPI** Cost Per Inquiry.

**CPM** Cost Per Thousand. Cost per thousand mailers sent or phone calls made. Also the cost per thousand people with the potential of seeing your advertisement.

**CPN** Calling Party's Number. The technical way of referring to the telephone number passed along with Caller ID services. The telephone number of the calling party. An AT&T spokesperson explains that this phone number is different than the one you get with ANI. With ANI you get the billing party's number, which in the case of a company using Centrex service, would be different from the actual number the call was made from. See CALLER ID and ANI.

**CPO** Cost Per Order.

**CREAMING** When you use or are given the most responsive part of a list for your test. This gives you an artificially high rate of response.

**CRM** See CUSTOMER RELATIONSHIP MANAGEMENT.

**55**

**CROSSTALK** When you (or one of your agents) hear another conversation on the telephone line — and it's not a conference call — you are experiencing <I>crosstalk. The phenomenon is caused by poorly shielded trunks or lines, certain transmission techniques and wire placements.

It's not ghosts, it's not a wiretap, and it shouldn't be confused with the telecommunications software program with the same name. The temporary solution? Break the connection and call again.

**CSC** 1. Customer Service Center 2. Customer Service Consultant 3. Customer Service Coordinator 4. Customer Support Center 5. Customer Support Consultant.

**CSR** Most likely, "customer service representative." In your call center is a CSR a customer service representative? Be careful when talking to your telecom people. CSR also means "Customer Station Rearrangement." When you have Centrex this is what they call changing station assignments.

To the telephone company it also means "Customer Service Record." This computer printout lists all the tariffed, fixed monthly charges on the bill from your local telephone company. Again, a very important term to telecom people.

Before you have a chat with your telecom manager or your local telephone company rep about CSRs, make sure everybody understands what is being discussed. See AGENT.

**CTI** Computer-Telephone Integration. Synonymous in most contexts with "computer telephony." A term for connecting a computer (single workstation or file server on a local area network) to a telephone switch and having the computer issue the switch commands to move calls around. The classic application for CTI is in call centers. Picture this: A call comes in. That call carries some form of caller ID or ANI. The switch "hears" the calling number, strips it off, sends it to the computer. The computer does a lookup, sends back the switch instructions on what to do with the call. The switch follows orders. It might send the call to a specialized agent or maybe just to the agent the caller dealt with last time. There are emerging standards for CTI. This is also called Computer Telephony, which is perhaps a more elegant term.

**CURSOR DIALING** See PREVIEW DIALING.

**CUSTOMER CARE** A much-talked about call center concept, which means simply this: not merely waiting for your customers to call, but offering excellent service by calling your customers. The concept has different applications depending on your business. Customer care can include follow-up calls that make sure your customers received the service or product they expected, continuity programs that offer something to your best customers, calling your customers when you expect them to need to refill an order ("Your probably down to your last box of paperclips") and excellent inbound service based on extensive customer databases.

**CUSTOMER CARE CENTER** A term created to describe a telephone call center with three basic elements: First, the database technology and the marketing savvy to fill that database with individual customer preference information. Second, the ability to intelligently handle inbound phone calls. Third, the ability to intelligently make outbound calls.

**CUSTOMER INFORMATION SYSTEMS** See CRM or CUSTOMER RELATIONSHIP MANAGEMENT.

**CUSTOMER PREMISES EQUIPMENT (CPE)** Terminal equipment — telephones, key systems, PBXs, modems, video conferencing devices, etc. — connected to the telephone network and residing on the customer's premises. The stuff — usually the phone stuff — that your company owns and keeps in its offices or facilities.

**CUSTOMER RELATIONSHIP MANAGEMENT** CRM. Customer relationship management is a way of tying the front end of the customer interaction with the vast resources of data on that customer that exist in the back (and usually outside the call center). From a customer point of view, that means that the agent information ready on who the caller is and what dealings they have had in the past. It results in shorter calls, and better calls that generate more revenue and are more refined and purposeful. It's only since CTI has become mature that this is really possible to accomplish on a large scale; until recently, this has been one of those apps available only to the largest call centers, usually as an extremely expensive, slow customization. All that is changing as the technology improves.

**CRM** has become one of those hot terms - every vendor wants to be a CRM vendor, and claims to have CRM tools. CRM has been co-opted by every company that pulls or pushes customer data in any direction. Also see E-CRM.

**CUSTOMER SENSITIVITY KNOWLEDGE BASE** A term created to describe a complex database that keeps track of your customers' preferences. Such a database would be updated almost automatically based on every contact you had with the customer. The database would probably be object-oriented since the idea is to define customer preferences based on individual preferences, not on a statistical analysis of conglomerate preferences such as those typically gleaned from existing character databases.

**CUTOVER** The date on which an ACD system goes online and handles live traffic.

**CVP** A British term. Co-operative Voice Processing, which gives the caller the ability to move seamlessly between an Interactive Voice Response system and a live agent.

**D CHANNEL** In ISDN, the "data" channel. See ISDN.

**DATA DIP** No, it's not that really weird guy in the MIS department. Data dip is an interactive voice response (IVR) term that means looking up information in a database using caller input or other caller information. In particular it means getting information on a caller (name, address) based on his or her CLID (caller ID) by looking up his or her telephone number (that's what caller ID gives you) in a big database. This big database is often external to your company: it is one of the same databases used for list look-up services.

**DATA MINING** Data mining is the analysis of data for relationships that have not previously been discovered. In effect, what data mining does for you is allow you to explore information that's been gathered as a result of long stretches of normal business operation, years or even decades, and discover things about your customers and products that you didn't know before. You might, for example, learn things from a close look at correlations between calling patterns and customer buying histories. Data mining is not something that goes on in call centers. However, call center data is a mother lode of information for data mining research.

**DATA RESERVOIR** If a data mine is the place where information resides in its rawest, ugliest, least analyzed state, and a data warehouse is where it lives once it's analyzed, what's a data reservoir? The ready pool of information, usually customer data, that's in use by a call center and is available at any moment during the customer interaction. Ahhh.

**DATA WAREHOUSING** A fancy way of saying "database", or a collection of databases. Especially when used for the rigorous analysis of data mining. Increasingly, this kind of concept (long a part of IT thinking) is being heard in and around the call center. This is because the call center is being integrated into corporate data structures, and IT is reasserting control over the hordes of information that the call center spits out. Until recently, call center data wasn't thoroughly analyzed for much beyond simple call center productivity. Now, since companies want to know how valuable certain customers are and how to most effectively sell to them, they are looking at call center data as a source of knowledge. And the best people to exploit that knowledge are the IT people who build data warehouses and then mine them.

**DATABASE INTEGRATION** Connecting your database to any of your other business systems, like your phone system, or ACD, or even a multi-user sales program.

The advantage of database integration with phones and ACDs is that when customers call, their information can be sent to the agent at the same time as the call. And when you share databases among the various departments that come in contact with customers, each person (whether in sales, customer service or technical support) has the same information about that caller's history and preferences.

**DATABASE LOOKUP** A software program function which gives you information about your caller by associating an ID number with a record in your existing database. This ID number may be a telephone number supplied by CLId (Calling Line Identification) or ANI (Automatic Number Identification). It can also be a telephone number, account number, social security number or other ID entered by the caller into an IVR system before the call is connected to an agent.

**DATABASE MANAGEMENT SYSTEM** A computer program (or programs) that lets you create, maintain and access a database. Some popular database management systems for PCs are dBase and Access. There are other database management systems for other types of computers. Don't confuse the database (the actual information or data on your computer) with the database management system (the software that controls it). A database management system is called DBMS for short.

Database marketing is an information funnel. Data about your customers comes in through your company's operations. If you are smart enough to store this info in a database, you can use it for better field sales and technical support, to create more targeted direct mail and more.

**DATABASE MARKETING** The art and science of designing a marketing campaign using information from your customer database. Many of these campaigns analyze traits a company's customers have in common, then try to "clone" these customers by appealing to other prospects with similar traits. Creating a demographic model can also help the company serve existing and prospective customers better by providing information about buying habits and personal preferences. It can also help in the development of value-added services.

For example, lets say you run a hunting catalog. By examining your database you find most of your customer are men between the ages of 35 and 60 who live in rural areas. These may lead you to buy a list from a football magazine popular with men between the ages of 35 and 60, and select rural zip codes for a catalog mailing or a telemarketing campaign. Your analysis may also show, that while a small number of the total, your fastest growing market is women between the ages of 20 and 40 in suburban areas. You may decide to buy a list from a children's clothing catalog that sports the same demographics. Or you may decide to exhibit at a gardening expo to reach out to this new market. In either case you may find your customers all have full time jobs, and decide to open your call center 24-hours a day to make it easier for them to order.

**DATABASE SERVICE MANAGEMENT INC** See DSMI.

**DAY-OF-WEEK FACTORS** A historical pattern consisting of seven factors, one for each day of the week, that defines the typical distribution of call arrival throughout the week. Each factor measures how far call volume on that day deviates from the average daily call volume.

**DBMS** See DATABASE MANAGEMENT SYSTEM.

**DCOM (DISTRIBUTED COMPONENT OBJECT MODEL)** Cousin of Microsoft's Component Object Model for software development. This modular object-oriented model lets programs request services from other programs on other computers on a network. It differs from COM in that COM describes a set of interfaces enabling communication within the same Win 95 or NT computer. See COM.

**DCS** Digital Crossconnect System. A device for switching and rearranging private line voice, private line analog data and T-1 lines. A DCS performs all the functions of a normal "switch," except that connections are typically set up in advance of when the circuits are to be switched — not together with the call. You make those "connections" by calling an attendant who makes them manually, or dialing in on a computer terminal that is similar to the one airline agents use.

**DDD** Direct Distance Dialing. AT&T's basic long distance service. Sometimes used as a generic term for no-frills long distance service, but it usually refers specifically to the AT&T service.

**DEALER LOCATOR** An application of ANI technology. It works can this: A caller who needs information about a product calls your 800 number. As the call is routed to the agent, the caller's phone number is compared with a database of your dealer locations. When the agent gets the call, he or she can refer the caller to the dealer closest to the caller's location. Or, the entire process can be automated and an IVR system can deliver the dealer locations.

**DECISION TREE** See BRANCHING SCRIPTS.

**DECOY** A name planted in a list to track the use of that list. It's a lot easier to use for mailing lists than calling lists. Usually the name is completely fake, or mangled in some way so the recipient knows, "OK, this is a response from that list." Unfortunately, many outbound telemarketing agents mangle peoples' names so badly, that it's harder to tell if the call is a mistaken "real" call, or a call generated

**61**

from the decoy name on the list. The record appears on that list only to assure accuracy. Also known as a seed or a dummy.

**DEDICATED** A device that has only one function. A PC on a network that is used only for printing would be a "dedicated server," for example. A dedicated line connects two points and does not carry other network traffic.

**DEDICATED ACCESS** A connection between a phone or phone system (like an ACD) and an IntereXchange Carrier (IXC) through a dedicated line. All calls over the line are automatically routed to a particular IXC.

**DE-DUPE** List biz slang for removing duplicate names from a list.

**DELAY ANNOUNCEMENT** A pre-recorded announcement that tells your callers that their call is being placed in your ACD queue. Just about everyone knows enough to say that the call will be answered soon by the next available operator. There are a number of other things you can do with these announcements to increase service to your customer. You can include in your recording the estimated wait-time. (And you can make it an excellent estimate with the right technology.) You can ask the caller to prepare information needed for the transaction, an account number, credit card or product code. On a technical support line, you can give your most frequently given tips ("make sure the device is plugged in...") or run down the solutions to frequently reported problems.

**DELAYED CALL STATISTICS** ACD statistics that show how long callers are willing to wait on hold before they are connected to an agent.

**DELUXE QUEUING** A feature that allows incoming calls from phone users, tie trunks and attendants to be placed in a queue when all routes for completing a particular call are busy. The queue can be either a Ringback Queue (RBQ)—the user hangs up and is called back when a trunk becomes available — or an Off-Hook Queue (OHQ) — the user waits off-hook and is connected to the next available trunk. Deluxe Queuing is a term used mainly by Lucent. Most modern PBXs have this feature. Most have simpler names, however.

**DEMAND DIALING** More sophisticated than "preview dialing," but not as complicated as "predictive dialing." One agent dials a call or enters a list of numbers to call, but is screened from the tones, busies and no-answers. When an answer is detected, the agent is connected. The calls are generated one after another. There is no attempt to anticipate when the agent will be finished with a call. The next call is dialed when the previous call is completed. Compare to PREVIEW DIALING and PREDICTIVE DIALING.

**DEMOGRAPHIC OVERLAY** A strategy to determine what people are likely to buy by taking known data, like a person's phone number and address, and assigning probable demographic characteristics to the person based on that. For example, you may assume a person's income is at a certain level because he lives within a certain

Zip Code. When used to carve up lists for specific campaigns, it can help find appropriate targets despite its relative imprecision.

**DIALBACK SECURITY** Dialback security is a telecom security feature. If a person calls to your telephone system for remote access, the system asks for a password. If the password is correct, the system hands up and dials back a pre-defined remote number. Only then does the caller receive access. Excellent security against hackers (doesn't do much for employee abuse though). It can be made even more secure with multiple passwords and features like voice recognition.

**DIALED NUMBER IDENTIFICATION SERVICE** See DNIS.

**DIALOG BOX** A feature of a software's graphical user interface. It is a prompt by the program for more information regarding the task you want accomplished. For example, you may ask a program to copy files — a dialog box will pop up and ask you which ones. Or in contact management software, you might want to sort your database, so you type S for Sort, and a box asks you for the name of the field you want the sort based on.

**DID** See DIRECT INWARD DIALING.

**DIGITAL ANNOUNCER** An electronic device that uses a computer chip to store a recorded announcement. In call centers that announcement will be played over telephone lines. Similar digital announcers are used for "don't park" and "no smoking" announcements in airports and sound effects in amusement parks. In call centers, digital announcers are frequently used for playing message-on-hold or music-on-hold recordings. Because they are digital, the recording can start from the beginning immediately after the end is played. There is no need to rewind. The announcement or program played on the announcer is often downloaded from a cassette tape recorded in a studio. Digital announcers' advantage over tape cassettes is that the sound does not degrade with repeated playings, there are no parts to wear out and you don't have to worry about a fragile, moving part (the tape) snapping in half. They are more expensive than tape players.

**DIGITAL COUPONING** Coupons used to be things that came in the Sunday papers... they still are, but now you can visit a Web site and download customized coupons for goods and services, print them out and take them to a retailer. The coupon is impressed with a bar code from the manufacturer that incorporates all sorts of data about the user and the kind of interaction that occurred online. From the manufacturer's point of view, this is better than traditional coupons because of the data tracking they can do, for the first time directly correlating a person's Web visits with retail transactions. For all their value to marketers, they have not yet caught on in a big way.

**DIGITAL CROSSCONNECT SYSTEM** See DCS.

**DIGITAL RECORDING** A way of recording that converts aural (sound) information into a series of pulses that are translated into a binary code intelligible to computer circuits. The information recorded this way can be stored in many ways, including as information in a computer.

**DIRECT DISTANCE DIALING** DDD. A telephone service which lets a user dial long distance calls directly to telephones outside the user's local service area without operator assistance. AT&T's name for its most basic long distance service.

**DIRECT INWARD DIALING** DID. A PBX feature that lets callers reach their party directly, without going through the system attendant (also known as the company receptionist). It eliminates the need for an automated attendant to route the calls after they arrive at the switch. A similar function of networked fax systems is also called DID.

**DIRECT INWARD STATION ACCESS** See DISA.

DIRECT MAIL A letter, kit or brochure sent through the mail to a business or residential consumer usually for the purpose of eliciting a prompt response of some kind from the consumer. Often this response is a telephone call, which is where a call center comes in.

**DIRECT MARKETING** An umbrella term that includes direct mail, telemarketing and direct response advertising. Any marketing that is aimed at a potential customer with the goal of having the customer respond directly to the marketer, and not to a retail establishment or other third party.

**DIRECT RESPONSE** Used to describe advertising. An ad that features a telephone number, order form or coupon designed for immediate response from a business or individual consumer. The response usually is placing an order for the product or service advertised. This term is also sometimes used to mean the same thing as "direct marketing."

**DIRECT-MAIL SOLD** A characteristic often noted on consumer lists for rent. Very important if you plan on doing a direct mail campaign, not as meaningful if you plan on telemarketing.

**DIRECTORY ASSISTANCE** DA. Formerly known as "Information", but changed because they were getting too many stupid questions. DA is provided by the local telephone company. In most states, the local phone company charges for this service. Most local phone companies will give you the person's address as well as his phone number if you ask for it. But you can usually only get two numbers per phone call (or sometimes just one). Directory assistance is an important resource for updated telephone lists and skip tracing. (People often move within the same city or town.) Directory assistance databases are now available on-line and on CD-ROM through major list and list processing vendors. Using these products can be less expensive and less time consuming than calling directory assistance for every two numbers you need to check.

**DIRECTTALK** DirectTalk is IBM's family of voice processing products, introduced in 1991. It is commonly used for interactive voice response (IVR) applications. There are two versions of DirectTalk: DirectTalk/2 which requires OS/2 Version 2, and DirectTalk/6000 which requires AIX and runs on any IBM RISC System/6000.

**DISA** Direct Inward Station Access. A telephone system (usually PBX) feature which lets outside callers call into the system, put in a password and dial out on the company's WATS or other special long distance service lines. It is most commonly used to let salespeople, or other traveling employees, take advantage of the company's special, low-cost long distance services.

A significant problem with this feature is that unauthorized people can break the password — with a computer program or simple guile — and run up your bill with long calls to foreign countries. Companies have lost thousands of dollars through this kind of fraud. If your phone system has DISA be strict with security or disable it altogether.

**DISCONNECT RATE** In an outbound environment, the percentage of calls a dialer makes that are dropped from the queue before reaching an agent — calls that end in busy signals, answering machines, or no answer. In an inbound center, the disconnect rate is the number of callers that reach the ACD, but hang up before they reach an agent.

**DISCRETE SPEECH** A type of speech recognition where the speaker must say each word separately, with a distinct pause between words. Sometimes a voice processing system will prompt the caller for the next word with a beep. See CONTINUOUS SPEECH.

**DISCRETIONARY PREVIEW DIALING** The agent presses a button to initiate dialing the next call. Compare this to forced preview dialing, where the next screen is brought up and the next telephone number is dialed as soon as the agent hangs up from the first call. See PREVIEW DIALING.

**DISTRIBUTED CALL CENTER** A center that is really several centers spread out, linked together, and managed as a single virtual entity. More properly known as a VIRTUAL CALL CENTER.

**DISTRIBUTION GROUP** A group of telephone extensions or agents on an automatic call distributor (ACD). Usually used to mean the group of extensions assigned to receive a certain type of call — determined by incoming trunk or other criteria. (The term begs the question of complex routing patterns.) Basically, the distribution group is the group of agents or extensions assigned to receive a particular call.

**DISTRIBUTION LIST** A voice mail or facsimile feature that is similar to a distribution list printed on a paper memo. A sales manager may have one distribution list of all her salespeople and another distribution list for upper management. For voice mail, the list is usually limited to internal extensions, fax lists usually have only outside fax numbers.

**DM** Direct marketing. This is any marketing that puts the seller and the customer together without an intermediary such as TV, radio or newspapers. It usually refers to telemarketing and direct mail, but it could also include everything from trade shows to the guys who hand out flyers on the street corner.

**DMA** Direct Marketing Association. A trade organization for the direct marketing industry (with a slight emphasis on direct mail companies), based in New York, NY.

**DN Directory Number**. What Nortel calls the number used for non-ACD calls on the its Meridian telephone system.

**DNIS** Dialed Number Identification Service. DNIS is a feature of toll-free and 900 lines that provides the phone number the caller dialed. The DNIS number can be provided in a number of ways, inband or out-of-band, ISDN or via a separate data channel. This is very helpful in call centers that answer calls for a number of businesses or product lines. Each business or product line has its own toll-free number. These calls terminate at a single automatic call distributor (ACD), but are routed to special call groups based on the number dialed. (Or, the agent may simply be prompted to the number the caller dialed either through a screen pop or whisper prompt, and handle the call accordingly.)

For example, suppose a call center handles calls for several chains of hotels. The toll-free number dialed (Perhaps 800-MARIOTT) triggers the combined computer-ACD system to send the Marriott script to an agent along with the call. The agent knows to answer, "Hello, and thank you for calling Marriott," instead of a general greeting or the name of another hotel chain handled by the center. The agent's reservation script includes information and specifications just for Marriott. Perhaps Marriott has a frequent- stayers club the agent should ask the caller about membership in.

DNIS is a boon to call centers that handle multiple product lines. The information relayed to the agent is completely transparent to the caller. As far as the caller is concerned, this agent or call center IS Marriott, and has no connection with another company or product. An alternative is to use an IVR system or an automated attendant to transfer calls to the correct ACD group or provide prompting, but using DNIS is a much more elegant solution.

**DO NOT CALL** The commonly used name for the Telephone Consumer Protection Act, an FCC regulation, effective in December 1992, that requires companies to keep a list of consumers who have requested not to receive phone solicitations from that company. Other provisions of this act say companies can't call consumers at home between 9 PM and 8 AM, companies must obtain consumer's consent to share his or her phone number with other marketers, and also outlines rules for ADRMP use.

**DO NOT CALL LIST** If your company makes outbound sales calls to consumers, it is required by federal regulation to keep a list of consumers who have requested not to receive telephone calls from your company. If your company calls someone who has requested not to be contacted, they can be fined. Exceptions to this regulation are non-profit organizations, market researchers and companies that have a prior business relationship with the person they are calling. (This includes collection calls.) On the state level, at least two states have "asterisk laws" that create a statewide do not call list for all companies. The Direct Marketing Association also has a list of consumers who have requested not to receive solicitations from companies. It is not illegal to call these people, but you can probably save yourself some time and money by skipping them. They've already made it clear they are not interested. See ASTERISK LAW.

**DOCUMENT ON DEMAND** A system that lets callers retrieve information by calling in to the system. In a call center, usually used to mean the same thing as "fax on demand." See FAX ON DEMAND.

**DONOR/CONTRIBUTOR LIST** People (or companies) who have donated money to a fundraising organization (such as charities, alumni groups and museums).

**DOUBLING DAY** Experience will show you that on certain day after a mailing you will have received 50% of all replies from that mailing. Only experience with a particular mailing and a particular audience can tell you which day it is though. That day is doubling day.

**DOWNLOAD** 1. The act of receiving a file or other packet of information that is being transmitted by another remote computer. 2. The file being downloaded.

**DSMI** Database Service Management Inc. The company that oversees the nationwide database of toll-free numbers and the software behind it. Day to day administration of the database is performed by a division of Lockheed. The physical database itself (actually, there are backups too) are located in Kansas City, Missouri. Some people in the toll-free number game pronounce this company's name "diz-me." The important thing to know is the company itself does not.

**DTMF** Dual Tone Multi-Frequency. Telecom lingo for touch-tone. These are the sounds your telephone makes when you press the buttons on the telephone keypad. A similar set of signals (called merely MF) are used inside the telephone network for signaling between the big network switches. DTMF is also used for data entry through IVR systems.

The tone you hear when you press a touch-tone key is actually two tones, one high frequency and one low frequency tone, transmitted at the same time. While there are 12 keys on the standard telephone, there are just seven tones emitted by the keypad. (There is another set of tones for special purposes, not generated by standard phones, which brings the total number to eight tones.) The "dual tone" concept works like this: all digits in the same row on the keypad share the same low frequency tone. All digits in the same column on the share the same high frequency tone. Each key has a unique combination of the two tones.

**DUAL HEADSET** Also known as an integrated headset. A special type of headset for the blind. One jack plugs into a telephone and another jack plugs into a specially configured PC. This PC provides voice synthesized output. The dual headset allows a visually impaired agent to work in a call center with no deterioration of service.

**DUAL-TONE MULTI-FREQUENCY** See DTMF.

**DUMB SWITCH** A word for a telecommunications switch that contains only basic switching software and relies on instructions sent to it by an outside computer. Those instructions are typically fed to the "dumb" switch through a cable from the computer to one or more RS-232 serial ports on the dumb switch. By itself, a dumb switch is just a generic telephone switch. Depending on how it is set up, and how the computer is programmed, it can become an ACD, a predictive dialer or other specialized switch. The dumb switch is a key tool in computer-telephone integration.

**DUMB TERMINAL** A computer terminal with no processing or no programming capabilities. It derives all its processing power from the computer it is attached to — typically over a local hardwired connection or a phone line. The good points: They're cheap. They're foolproof. Operators don't have to mess with floppy disks. They require minimal training. Their disadvantage is that everything must come from the central computer — not only the information (data record) but also the form in which to put it. The current trend toward client/server computer system architecture has made dumb terminals unfashionable.

**DUPE** Duplicate.

**DYNAMIC ANSWER** The ability of an automatic call distributor (ACD) to change the number of rings before the call is picked up by the switch to match the length of the wait time (or queue period). This is beneficial because it cuts the length of your toll-free (you pay) calls, so you only have to pay for the talk time. Hold time is eliminated is cut considerably. The problem is your callers may think you are not answering, and hang up. They might be so annoyed by you not answering that they never call again.

**DYNAMIC INBOUND/OUTBOUND** Refers to a features of Melita International's predictive call processing system that manages both inbound and outbound calls automatically for each agent. Agent resources are then allocated so that inbound call demand is met and given priority over outbound calls.

**DYNAMIC LOAD BALANCING** A sophisticated technique for routing calls, especially among multiple ACD systems. With dynamic load balancing the system isn't stuck looking for an available agent in the second or third hunt group. It can go back to the first group if an agent becomes available.

**DYNAMIC OVERLOAD CONTROL** DOC. The feature of a telephone switch which uses its translation tables and intelligence to let the switch adapt to changes in traffic loads by re-routing and blocking call attempts.

**E-MAIL** Electronic mail. A method of sending messages in the form of electronic text files from one person to another through a communications network.

**E-COMMERCE** Electronic commerce. Any kind of transaction that occurs in an on-line or telephonic environment. And we use this term to include information transactions, as when someone checks a bank balance by Web, or gets stock quotes by phone. See E-TAILING.

**E-CRM** This is an offshoot of CRM, or customer relationship management. In this case, it extends the idea of the customer interaction past the phone call, to incorporate information gathered from emails, website visits, live-chat sessions, and any other non-traditional non-telephony channels.

**E-TAILING** Electronic commerce refers to any form of fulfilled transaction that occurs in a "virtual" environment — everything from over the phone to via the internet. There seems to be some disagreement over whether the term refers to actual transactions that occur, where money changes hands, or whether it can be taken to mean a more all-encompassing information exchange, as when a Web surfer shops for something by Internet, and then performs the actual purchase through more traditional means. E-tailing, by contrast, is a subset of this that refers to actual on-line selling of goods, along the lines of an Amazon.Com or CD-Now. The thing that bugs us about it is that the "e" in e-tailing doesn't stand for anything, as it does in e-commerce or e-mail. It's there just to make you think of retailing. Is there anything more infuriating than an unnecessary synonym for e-commerce, itself a silly term? (Describe late 20th century or early 21st century e-commerce to a banker, and he'll tell you that's what his industry has been doing for the last 25 years.)

**EAR LOOP** A headset term. The oval-shaped, plastic piece used to secure a headset's earpiece to the user's ear. It is only used on monaural (one ear) headsets. These days it is often part of a kit that includes also a headband and behind-the-ear stabilizer. The headset is convertible between the three styles. The ear loop is the preferred headset style for someone who feels constricted by a headband (or doesn't want to get his or her hair mashed by the headband), but finds the behind-the-ear style not secure enough. The ear loop is often recommended for glasses wearers.

**EAR RING** An oval-shaped, plastic ring used on a headset to secure the headset's earpiece to the user's ear. See EAR LOOP.

**EAR TIP** A very small headset earpiece that actually fits inside the outer ear. It provides superior sound quality, but some people find it uncomfortable.

**ECH** Enhanced Call Handling. Systems that handle telephone calls "intelligently" through a variety of network, human, computer and telecommunications resources. ECH systems include voice mail, interactive voice response, computer-telephone integration, fax-on-demand and complex telephone networks.

**ECHO SUPPRESSOR** An echo suppressor is a device used on telephone connections. It cuts down on the echo in a voice transmission by making the circuit one way — by turning off transmission in the reverse direction while one person is talking. Echoes are particularly prevalent and annoying on satellite circuits.

But this one-way transmission ticket does full-duplex data transmission no good. An echo suppressor can be turned off by a high-pitched tone generated by an answering modem. This tone tells the circuit that the "conversation" will consist of data, not voice.

**ECP** See ERROR-CORRECTING PROTOCOL.

**EDA** Electronic Directory Assistance. A directory assistance database stored on CD-ROM or in a computer database accessible commercially through an on-line service. EDA is an often less-expensive alternative to multiple telephone calls to a telephone company's directory assistance number. It is very helpful for updated telephone calling lists and in collections departments for skip-tracing.

**EIGHT HUNDRED SERVICE** See 800 SERVICE.

**ELECTROMAGNETIC RADIATION** See EMR.

**ELECTRONIC CALL DISTRIBUTION** Another term for Automatic Call Distribution. See AUTOMATIC CALL DISTRIBUTOR.

**ELECTRONIC COMMERCE** See E-COMMERCE

**ELECTRONIC-COMPATIBLE** A headset that works with phones that use an electronic microphone as a receiver. This type of headset (and telephone) requires an amplifier and an additional power source.

**ELECTRONIC HOMESTEAD** Telecommuting as stretched to its logical end. The use of networks (data and telecom) to create a workstyle that echoes the agricultural homestead of the pre-Industrial age. Both Mom and Dad work at home, perhaps each at a number of part-time jobs. The most skilled workers and professionals will have the most flexibility and independence. Their "product" will be their skills and services. Less skilled or ambitious workers will mostly sell their time, and will primarily be employees, although they may work for more than one company.

**ELF** Extremely Low Frequency magnetic radiation. A slice of the electromagnetic spectrum emitted by video display terminals (AKA computer screens) whose long-term danger to human beings has been neither proven or disproven. Emissions are greatest to the rear and sides of the display. Something to keep in mind when designing the layout of your call center.

**EMR** Electromagnetic radiation. Electromagnetic radiation consists of a spectrum of energy at various wavelengths, from gamma rays and X-rays at very high frequencies to visible light in the middle, to microwaves and radio frequencies at the low frequencies. Many slices of the electromagnetic spectrum are important media for telecommunications, including visual light in fiber optic transmissions, microwaves for short-haul connections and radio frequencies for wireless and cellular communications.

**END-OF-SHIFT ROUTING** An ACD term for a process that assures calls won't be left in limbo when a shift ends.

**ENDLESS LOOP** A term used to describe a tape, a cassette or a tape player. Before digital announcers became cheap, the most popular way to achieve continuous play on a message-on-hold or music-on-hold recording was through a recording tape, set up, not with two ends, like the cassettes you play at home, but with the two ends connected to each other. No need to rewind. The thing just starts over when it's done. The disadvantage of an endless loop recording is it is prone to wear, tear and jams. The tape needs to be replaced periodically. It has the advantage over digital announcers in that it is still less expensive.

**ENHANCED DNIS** Enhanced DNIS is a combination of ANI and DNIS delivered before the first ring on a T-1 span. The number of digits delivered is configurable on a per span basis.

**E911** A telecommunications service that provides emergency service dispatchers with information (name, address, a map) that is linked to the calling telephone number. So, if someone calls because his house is on fire, but drops the phone and runs out the door before saying anything, the dispatcher still knows where that person was calling from, and can give the fire department directions for finding the house.

E911 service is important to the call center industry for several reasons. First, it provided "screen pops" before ANI or CLID were available to the public and long before CTI was a buzzword. Second, E911 calls are answered in call centers, centralized locations answering high volumes of calls. Yet, these dispatch centers do not think of themselves as part of the call center industry.

**ERGONOMICS** The science of determining proper relations between mechanical and computerized devices and personal comfort and convenience; that is, how a telephone handset should be shaped, how a keyboard should be laid out. Call

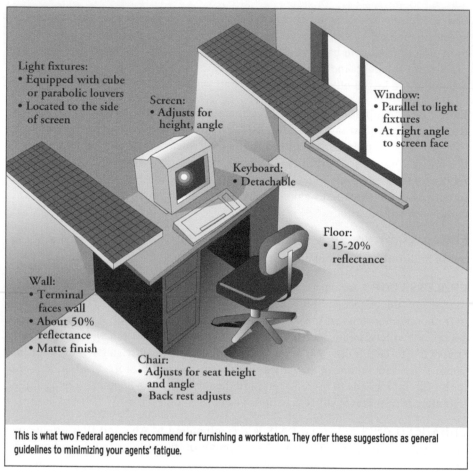

Light fixtures:
- Equipped with cube or parabolic louvers
- Located to the side of screen

Screen:
- Adjusts for height, angle

Window:
- Parallel to light fixtures
- At right angle to screen face

Keyboard:
- Detachable

Floor:
- 15-20% reflectance

Wall:
- Terminal faces wall
- About 50% reflectance
- Matte finish

Chair:
- Adjusts for seat height and angle
- Back rest adjusts

This is what two Federal agencies recommend for furnishing a workstation. They offer these suggestions as general guidelines to minimizing your agents' fatigue.

center employees (especially agents) have a particularly intimate relationship with the technology required to do their job. They also don't have much freedom in how and when they interact with that technology. (They must be at their stations during their entire shift, office equipment is bought in bulk to fit the average worker, etc.) For these two reasons ergonomics is of particular importance in call centers.

**ERLANG** A measurement of telephone traffic. One Erlang equals one full hour of use (conversation), or 60 x 60 = 3,600 seconds of phone conversation. You convert CCS (hundred call seconds) into Erlangs by multiplying by 100 and then dividing by 3,600 (or, dividing by 36). Numerically, traffic in a call center, when measured in Erlangs, is equal to the average number of trunks that are busy during the hour in question. Thus, if call center traffic is 12.35 Erlangs during an hour, a little more than 12 trunks were busy, on the average. Similar calculations can give you a rough number of the agents needed.

Any discussion of Erlang, the measurement, quickly evolves into a discussion of the Erlang formulas of queuing theory. Lucky for us, that's the next entry. See ERLANG FORMULA.

**ERLANG FORMULA** A mathematical way of making predictions about randomly arriving work-load (such as telephone calls) based on known information (such as average call duration). Erlang formulas are used to determine call center staffing (humans) and the number of trunks required.

Central to queuing theory are some basic facts. First, traffic varies widely. Second, anyone who designs a telephone switch to handle all peak traffic will find the switch idle most of the time. He will also find he's built a very expensive switch. Third, it is possible, with varying degrees of certainty to predict upcoming "busy" periods.

There are two types of Erlang formulas. Erlang B is used when traffic is random and there is no queuing. Erlang C is used when traffic is random and there is queuing. It assumes that all callers will wait indefinitely to get through. Therefore offered traffic (see ERLANG) cannot be bigger than the number of trunks available (if it is, more traffic will come in than goes out, and queue delay will become infinite). See also POISSON DISTRIBUTION and POISSON PROCESS.

**ERROR-CORRECTING PROTOCOL** A set of data communications rules (see PROTOCOL) that monitor file transmissions for errors. ECP's break down the file into blocks of various sizes. The integrity of each block is checked on the receiving end, which asks for a retransmission if an error is found. Error correction is important when using standard phone lines for data transmission because of the tendency of line-noise to leak in and corrupt the data.

**ETHERNET** A local area network system that operates over twisted wire and coaxial cable at up to 10 megabits per second.

**EVENT CODE** A code that an agent centers at the conclusion of a call. Event codes can trigger a variety of follow-up activities such as an acknowledgement letter or inclusion in a list for a subsequent campaign.

**EXCHANGE** The first three digits of a local telephone number. All telephone numbers with the same exchange are served by the same central office. With the right software (or a very large list) you can find out a company's location (city, state, maybe even which part of the city) if you know the area code and exchange.

**EXPERT SYSTEM** A sophisticated computer program with three parts. 1. A stock of rules or general statements. These rules are generally based on the collective wisdom of human "experts." 2. A set of particular facts. 3. A "logical engine" which applies facts to rules to reach all the conclusions that can be drawn from them. The idea of expert systems is to help people solve problems. In call centers, expert systems are most often the tools of help desks, which use them to solve customer problems. With an excellent expert system it is even possible to have novice help desk agents solve advanced or difficult problems.

**EXTENDED CALL MANAGEMENT** A Nortel term for a collection of features for its DMS Meridian central office Automatic Call Distribution (ACD) service. Using Switch-to-Computer Applications Interface (SCAI), ECM works with user-provided computer equipment to integrate call processing, voice processing (recorded announcements, voice mail and voice response) and data processing. For example, ECM lets an outboard computer device coordinate the presentation of customer data on the ACD agent's computer screen with an incoming call. The D channel of an ISDN Basic Rate Interface (BRI) serves as the transport mechanism from the central office switch to an outboard computing device. Communication is peer-to-peer, meaning that neither the switch or the computer is in a "slave" relationship to the other.

**EXTREMELY LOW FREQUENCY MAGNETIC RADIATION** See ELF.

**FACSIMILE MACHINE** The proper name for a fax machine. See FAX MACHINE and other FAX entries.

**FALSE RINGING** A recording of a telephone ringing signal which is played while a call is transferred or while a switching device listens for modem or facsimile mating calls.

**FAULT TOLERANT** The ability of a piece of equipment to keep running when it encounters a hardware failure. This is done through building in multiple redundancies at every critical point. If a component fails, there is supposed to be another one to take over without any loss of function or data.

**FAX BOARD** A printed circuit board with a fax machine's dialing and image processing circuitry on it, that fits into an empty expansion slot in a personal computer. Once merely a personal productivity gizmo, fax board can be combined with voice processing boards and software to create sophisticated fax-on-demand systems, IVR systems with fax and other applications.

**FAX CARD** Another term for FAX BOARD.

**FAX GATEWAY** A bridge between two processing systems (such two LANs, a LAN and a mainframe, or a LAN and the outside world) that lets individuals on either side to send faxes through to the other. Fax gateways are particularly helpful in call centers that send sales letters, appointment confirmations, order confirmations or product information by fax. The gateway lets all your agents access fax-sending capabilities from various networks. It gives you the advantages of a fax at each agent station without the cost.

**FAX MACHINE** A device that sends or receives a copy (a "facsimile") of printed material to or from a remote machine over standard phone lines. Fax machines are now so common in offices of all types that its hard to justify a lengthy description here.

Call centers should know a few things about sending faxes. First, federal law states that every fax sent must have the fax number of the sending fax on it somewhere. It can be on the cover sheet, or on the fax ID that is printed on the top of each page on the receiving end. Second, the "do not call" list laws apply to faxes too. If a company asks you not to send them promotional faxes, you can be penalized for sending them promotional faxes in the future. Sending junk faxes is a no-no. You should have prior contact with any company or person you send a fax to. It is always good form to send unsolicited faxes after working hours, when the company's fax machine is less likely to be in use.

Should you buy a regular fax or a plain paper fax? Faxes printed on the standard, slimy fax paper fade in time. And they black out when exposed to heat (left on a radiator or in the sun). If the fax is important, photocopy it immediately. Plain paper fax machines tend to jam a lot more than ones that use rolls of slimy paper. For some technical information, see GROUP 1, 2, 3, 4. Also see JUNK FAX.

**FAX MAILBOX** The fax equivalent of a voice mail system. Each person in an office is assigned a mailbox number. An outside person who wants to send a fax calls up the main fax number, and is asked by a voice processing system to input the mailbox number of the recipient. The fax is stored electronically in the "box" until the recipient retrieves it. Also, some systems let you retrieve faxes from your mailbox from outside fax machines. A similar application is possible by designating a mailbox on a voice mail system for receiving faxes.

**FAX ON DEMAND** The remote retrieval of archived documents through a fax machine. This is used a lot as a front-end for technical support call centers, wherein an IVR system will help a customer get diagrams or other complex information delivered to their fax machines. It's a lot cheaper to arm a customer with printed data, delivered instantly, than it is to have a first line agent trying to describe that information.

**FAX-OVER-IP** For all the hype over the promise of internet telephony, one very practical, nearly overlooked application is fax-over-IP, which is using the Internet to transmit fax traffic instead of the regular phone network.

When you send a regular fax, your fax machine goes off-hook, dials, and the phone network completes a circuit over phone lines to the receiving fax. You pay for the circuit. The two faxes negotiate a connection and exchange image data. When you send a fax over IP, however, your fax machine (or PC client) transmits the image data as packets through an IP data network — the public Internet, a company intranet, or an external extranet — instead of the phone company's network.

Fax over IP servers only need the phone network (if they need it at all) for the local legs of calls: to deliver to faxes to regular fax machines, or receive from regular faxes. Thus, you save the long distance costs - and it's free if it's inside your company. Faxing over IP you don't pay more than you ordinarily would to maintain your IP network connection. At most you pay the cost of the local call.

What's great about fax-over-IP, as opposed to voice, is that none of the irritating latency and sound-quality issues crop up. It doesn't matter if there are some errors, or if some of the packets arrive out of sequence, as long as the fax arrives.

**FAX SERVER** A computer with one or more fax boards installed and hooked up to a local area network whose primary purpose is to act as a fax station for all the users on the network. It sends faxes from any PC on the net, as well as receive them and print them out on a dedicated laser printer. Many fax servers also include voice boards that add fax-on-demand capability for people calling from the outside.

**FAX STORE AND FORWARD** The ability to have your faxes held by the receiving fax processor in electronic storage (as a file on a hard disk), and forwarded to whatever fax machine you'll be near, like one in a hotel or a customer's office. Kind of voice mail for faxes. See VOICE MAIL

**FCC** Federal Communications Commission. The federal agency that regulates interstate communications including radio, television — and most important to call centers — telecommunications. Instate telecommunications are regulated by state public utilities commissions.

The FCC is located in Washington, DC and is run by a seven member board appointed by the President. It basically does three things: 1. It sets the prices for interstate phone, data and video service. 2. It determines who can or cannot get into the business of providing telecommunications service or equipment in the United States. 3. It determines the electrical and physical standards for telecommunications equipment.

With recent moves toward government downsizing and deregulation, many of the FCC's traditional duties are being farmed out to industry committees. Your opinion on upcoming rulings is solicited through a "Request for Comment." (You'll only know about them if you pay attention.) FCC rulings can be appealed through a Federal Court.

**FEATURE BUTTONS** In all telephone systems there are codes that can be dialed to access a system feature. For example, dial *34 to put a call on hold. Think of a feature button on a telephone as a collection of numbers stored in a bin. When you hit the button, the bin quickly disgorges all the numbers one after another. It dials the feature code for you. In computer terms, a feature button on a phone is the same as a macro — an easy way of doing something. On most phones with feature buttons, the feature buttons are "programmable." This means you can assign different features to different buttons, i.e. the ones you want.

**FEDERAL COMMUNICATIONS COMMISSION** See FCC.

**FILTER** A database criteria that suppresses certain records. For example, looking at your customer database, you may only want to see your customers who live in Massachusetts. You would create a filter to screen out those with any other entry in the state field and show you only those in Massachusetts. A filter is used for creating subsets of the main database for examination and reports. It hides the unwanted records, but does not delete them. Filters are a feature of contact management software, sales automation software and predictive dialing software.

**FIRST-TIME BUYER** If you bothered to look this up, you probably saw it as a selection on a direct mail or telephone list. It usually means someone who has ordered something from the company for the very first time — not necessarily that this is the first time they have bought this type of product. The selling point is that this is a person who has not appeared on the company's lists before. New blood.

**FLASH HOOK** Another name for switchhook or hookswitch. The little button on the telephone that you place your receiver into. It hangs the phone up, releasing that line to receive another call. If you push the flash hook quickly, you can signal the switch at the other end (central office or PBX) to do something, such as place a call on hold and switch to the incoming one (call waiting), or transfer the call to another phone. Many feature telephones have a "flash" button, which performs that fast flash hook thing, relieving the user from the worry that they will hang up instead of switch to the other call.

**FLEXAGENTS** A term for agents who sign onto their predictive dialing system as both inbound and outbound agents. Dependent on call volumes, they can either take ACD inbound calls or be automatically reassigned to a predictive outbound call campaign. Similar to UNIVERSAL AGENT.

**FOCUS GROUP** A set of consumers, chosen on the basis of particular demographic characteristics, who are asked to respond to questions about products and services in a subjective, conversational, group-oriented manner. Their discussions are taped and analyzed and in some cases give a better sense of how people feel than a more objective survey.

**FOLLOW THE SUN DIALING** A technique used in call centers whereby the agents call those parts of the country where it's convenient to call and move the calling across the country as the sun moves. Agents call New York households between 6 P.M. and 8 P.M. When the time hits 8 P.M., the agents stop calling New York households and start focusing calls on households in the central time zone. To accomplish Follow The Sun Dialing, a call center needs software which knows in which phone numbers are in which time zones.

**FORCE FEED** A method of dialing calls that automatically dials and delivers a call to an agent a pre-determined time after the agent finishes the previous called.

**FORCED AVAILABLE** An ACD feature which marks an agent as available immediately when a call ends.

**FORCED PREVIEW DIALING** Instead of pressing computer key to dial the next call, the next computer screen is automatically brought up and the next number is automatically dialed when the agent hangs up the first call. See PREVIEW DIALING.

**FORECASTING** Taking historical data (what happened in the past) from your ACD and using that information to predict what might happen in the future. Has your call volume always doubled on Tuesday? It will probably double next Tuesday too. A very important function of call center management software.

**FOREIGN EXCHANGE** Also known as FX. If you have foreign exchange service, a caller in another city dials a local telephone number. The call is transferred to your headquarters in another city. The caller only pays for the local call. (You pay for the call from that point to your headquarters.) The "foreign" here means foreign to your central office (CO) not necessarily in another country. The CO is treating a call from another exchange (or city) like a local call. Airlines and rental car agencies seem to be fond of this service. It gives your company a "presence" in another city.

**FORKLIFT UPGRADE** Most hardware, an ACD for example, can be expanded incrementally. First you buy a cabinet that holds more cards (for trunks and extensions) than you need at that moment. As your company grows, you first upgrade the system by buying more cards for the empty slots. When the cabinet is full you buy an expansion cabinet. There is some hardware that does not expand however. You buy one configuration and that's it. When your company gets bigger, you need a forklift upgrade. The come and take the old one away and bring the new one in, presumably with a forklift. Your system can also be subject to a forklift upgrade when you already have expanded it as far as it goes or when it gets so old you can no longer get parts or support for it.

**FORMAL CALL CENTER** A British term. A call center or other business-by-phone operation where all of the staff are dedicated to telephone based work. See also INFORMAL CALL CENTER.

**FOUR-WIRE CIRCUITS** Telephone lines using two wires for transmitting and two wires for receiving, i.e., four total. Four-wire circuits offer much higher quality. All long distance circuits are four-wire. Almost all local phone lines are two wire. All analog phones are two-wire.

**FREE LIST** ACD-talk for the list or table of agents available and which of those agents have been available the longest.

**FREEPHONE** A British name for what we in North America know as toll-free service. It means you can have your customers call you for free on phone lines with special exchange codes. It's a British Telecom toll-free service in which the caller dials the operator and asks for "Freefone XYZ."

**FREQUENCY** How frequently a direct response customer buys (based on experience over a length of time). This information is important to track on your own lists and valuable when available on rented lists. Also: the rate at which a wave alternates. For our purposes (telecommunication signals and electrical currents) it is usually measured in Hertz (cycles per second).

**FRONT-END** A device or technology that comes first, before the main application. For example, ACDs (automatic call distributors) are often front-ended by some kind of announcement system (VRU). When you call for airline reservations you are often greeted by a recorded message asking you to have the date and time of your flight ready. That's the announcement system. When your call finally goes through (routed to the next available operator by the ACD) and you speak to an agent, that's the main application.

This concept applies to many technologies. A complex computer system may be front-ended by an easy-to-use graphical interface. An in some states a recorded sales message played over the telephone must be "front-ended" by a live operator who tells you a recorded message will be played.

**FTE** Full Time Equivalent. A way of measuring staff levels, especially for budgets and scheduling. It simply means the number of staff-hours required has been divided as though each body is working a full-time schedule. It tells you what your staffing needs would be if your needs were covered only by full-time agents. The hours may actually be covered by part-timers. (And sometimes it seems like they usually are.)

**FULL TIME EQUIVALENT** See FTE.

**FUNCTIONAL SPLIT** A division within an automatic call distributor (ACD) which allows incoming calls to be directed from a specific group of trunks to a specific group of agents.

**FUNCTIONALITY** A fancy way to describe what a product can do. When a manufacturer brags about her product's "increased functionality" it just means it can do more things than it could do in the last version of the product. It used to slice, now it slices, dices and chops.

**FUZZY LOGIC** Once upon a time, computers were annoyingly exact. They couldn't tell that Chuck Jones and Charles P. Jones were in fact the same person or that if you wanted to fly to Paris this afternoon, that a flight at 6 PM was better than none at all. These things required a human. But humans figured out how to program computers so they could recognize that things were similar, but not exact. This programming is called fuzzy logic and it has many applications in many industries.

In call centers it is most often used in problem resolution software, where it turns up solutions to problems similar to the one entered, even if there are no exact matches. This can mean the software will turn up a good solution even if a word in the problem entered is spelled incorrectly, or if there are different ways to phrase the problem. ("DeskJet drops columns" or "Table doesn't print correctly.")

Fuzzy logic is also used in call centers to find telephone listings for customers and for schedules of all kinds, from the airline example (which could be used in an IVR system) to creating schedules for call center agents (who might get something close to, but not exactly what they requested).

Fuzzy logic is a type of artificial intelligence.

**FX** See FOREIGN EXCHANGE.

**G-STYLE HANDSET** A telephone handset either has a round mouthpiece and ear-piece or square ones. If your phone has round pieces, it has a G-style handset. Most phones these days come with square handsets, known as K-style. There are a few things you can tell about your telephone and telephone system by knowing it has a G-style handset. First, it's probably pretty old. Second, it probably has a carbon-based microphone. (This is vital to know when you buy a headset.) Third, the old trick of tapping your mouthpiece on the desk when the other party can't hear you clearly actually works with G-style handsets. (Don't try it with K-style handsets!) It has something to do with lining up the carbon granules.

**GAB LINE** A 900 or pay-per-call service. See GROUP ACCESS BRIDGING.

**GALAXY** The brand name of Rockwell's signature automatic call distributor. It is no longer manufactured by Rockwell, but the company still supports those systems still in use.

**GANTT CHART** A Gantt chart is a visual display format used in call center man-agement and planning. Gantt charts, named for their inventor, came to call centers from the engineering world, where they've been used for decades. A Gantt chart shows, using rows and columns, a table of available resources spread out over time. They are used for project planning in other industries. In call centers, it's one way to view the availability of agents over increments of a staffing day.

**GATE** Also known as a split or a group, the term "gate" usually refers to a set of agents served by an ACD. In the ACD's routing scheme, a gate is a group of agents that are all qualified to handle a certain type of call. A call center may operate with a single gate, meaning any call could go to any agent. In larger centers, multiple gates are used.

Typically, a center will have a gate for each of its functions, one for customer ser-vice calls, another for orders, a third for technical support. Calls are routed by the number dialed (DNIS) or after the caller selects a choice from a voice response unit or automated attendant menu. Usually each gate has its own queue. In centers with more than one gate, and the technology to do so, there is usually an "over-flow gate" or group assigned to back up the primary gate if all agents are busy. A good ACD will let the call go to an agent in the primary gate, if that is the first agent to become available.

Rockwell offers this explanation: "A gate consists of trunks that receive a particular type of call and a group of positions capable of responding to that call. In the basic Rockwell Galaxy ACD system, for example, 32 gates are provided, but the expansion feature allows a maximum of 128 gates to be configured. The number of trunks and positions which can be assigned to each gate is restricted only by the system size limitation."

**GATE ASSIGNMENTS** An ACD gate is made up of trunks that require similar agent processing. Individual agents can be reassigned from one gate to another gate by call center management through the supervisory control and display station. In some ACDs, gate assignments are static. They are difficult to change. In others, its simple to create new gate assignments on the fly as conditions in the center change.

**GATEWAY** A link between networks. When two networks don't speak the same "language," that is, they don't use the same protocols, a gateway is used to convert communication between network into the correct protocol.

The same term is used to describe links between different computer networks (to link a LAN to a mainframe) or to link a telecommunications network and a computer network (a PBX or ACD to a mainframe or other computer system).

**GBH** Group Busy Hour.

**GEOGRAPHICALLY-VARIABLE OPERATING COSTS** Those costs associated with running a call center that change depending on where you locate your center. Labor, for example, costs less in the Midwest and the South than it does in the big cities of the East and West. Another variable cost is telecom service itself — the farther you locate your center from a point of presence on the network, the more it will cost you in long distance service.

**GOE** Grade of Efficiency. A term to define the percentage of time an agent is actually handling a call. It refers to the rate of calls, expressed as a percentage, for which agent connect time does not exceed a pre-defined threshold. Very close to what is called an OCCUPANCY RATE.

**GOS** See GRADE OF SERVICE

**GRADE OF SERVICE** GOS. A traffic engineering term that indicates the likelihood that an attempted call will receive a busy signal. Grade of service varies by time. It's obviously going to be worse during your busy hour than during a slack period. GOS is expressed as a decimal fraction. For example P.01 Grade of Service means an attempted call will have a 1% chance of reaching a busy signal. See TRAFFIC ENGINEERING.

**GREETING ONLY MAILBOXES** Mailboxes that deliver a message to incoming callers but do not allow a message to be left. The Greeting Only mailbox may transfer a caller to a designated telephone number.

**GROUP** For our purposes, "group" means a set of agents on an ACD. It's also known as a split or gate. It's a division of the call center for routing purposes. A certain type of call, perhaps arriving on a set of trunks, is delivered to the set of agents designated to receive those calls. See GATE for more.

"Group" also has a Signal Computing System Architecture (SCSA) meaning, which may come up if your call center creates its own voice applications. A group combines the functions of one or more Resource Objects with defined connectivity.

**GROUP 1, 2, 3, 3 BIS & 4** All are international facsimile standards. Group 1 and 2 are unbelievably slow. They are used only on the oldest machines. A Group 3 machine sends information at 9600 bits per second, or around 20 seconds for each standard 8.5 by 11 inch page. Most fax machines sold today are Group 3 compatible. Group 3 BIS is a fax communication standard related to the regular Group 3 protocol that goes as fast as 14.4 kilobits per second. It's used for high-resolution image transmission.

Group 4 faxes are transmitted at 64 kilobits per second, using the B channels on ISDN lines. (Regular lines just don't have the bandwidth to transmit this fast.) It transmits a page in about 6 seconds. Group 4 is used in special applications where the need for a high level of detail makes the higher cost worth while.

**GROUP ACCESS BRIDGING (GAB)** A teleconferencing service that allows many people to listen and speak with one another over regular telephone lines. In business it is used for teleconferenced meetings and presentations. The consumer application of this technology is for those "party lines" that gave 900 numbers a bad name. This is why they were called "gab lines."

**GROUP BUSY HOUR** GBH. The busy hour on a given group of trunks, or the busy hour for a particular group of agents. See GROUP

**GROUP HUNTING** A feature which automatically finds free telephones (or agents) in a designated group. See HUNT GROUP

**GROUP SCHEDULING SOFTWARE** A type of software that creates a schedule, not only for workers, but also for meetings, projects and office resources such as conference rooms. Similar, but not identical to the very popular "scheduling software," which creates the staff schedules for most call centers. See SCHEDULING SOFTWARE

**GTN** Global Transaction Network. An AT&T service that adds smarts to the routing of inbound toll-free calls. It lets you route incoming toll-free calls between centers in different geographic locations. It offers six call processing services: Next available agent routing, call recognition routing, transfer connect service (allows agents to transfer calls to distant ACDs), network queuing, toll-free select again service and multiple number database (allows multiple toll-free numbers to be assigned to a single routing plan in the network, rather than each toll-free number having its own unique routing plan).

**GUI** Sometimes pronounced "gooey." See GRAPHICAL USER INTERFACE.

**GUIDE** In IMA's EDGE telemarketing software, a series of screens that, when linked together, form a script through which you perform telemarketing or other call center activities. A guide has several branches or paths that can be taken.

**H.320** A video compression format that provides video conferencing over one or more 64 Kbps channels. Those channels are usually delivered by ISDN. You can think of H.320 as the ISDN video conferencing standard. You will need to think about video compression formats if you plan to offer your customers video access to your agents.

**H.323** A video compression format that provides video conference over the Internet. Some people will tell you that the quality is almost as good as the H.320 standard. Others will tell you it doesn't come close. (It probably depends on what they're sending and the quality of the Internet connection.) H.323 has caused some call centers to take a second look at providing video access to their call centers.

**HACKER** In common usage, this is someone who hacks his or her way into a computer or telephone system by illegal means. Use this meaning when talking to a computer programmer of a certain age, and you will see him or her turn purple with rage. Once upon a time "hacker" was a person who hacked away at a program until it worked. It then became a badge of honor for elite programmers. But common usage has taken over the meaning from the in-group term. If someone is referred to as a "hacker" you can bet he or she is breaking into to networks, not creating software, unless the context tells you otherwise.

**HANDLED CALL** A call that is answered by an employee, as opposed to being blocked or abandoned.

**HANDOFF** An SCSA definition. The change of ownership of a Group (and therefore, typically, a call) from one session to another. For example, if a call center application discovers that a caller wishes to access a technical support audiotext database, it hands off the call to an application servicing that database.

**HANDSHAKING TONES** That annoying screech you hear when you mistakenly dial a fax number with a voice phone is a fax machine trying to find out if it's talking to another machine. Unless you respond with an appropriate screech that tells it what kind of machine you are, what communications protocol and speed to use, no transfer will take place. The process of negotiating the parameters of transfer is called the "handshaking."

**HARD DOLLAR SAVINGS** For our purposes, the money you save by buying a product. Hard dollar savings are those savings that directly affect the bottom line. Lowering your telephone bill, reducing overtime or eliminating entire job positions are hard dollar savings.

Soft dollar savings are less tangible: your customers like your company better, your employees have more time to do a better job. They're nice, but you can't put them on a balance sheet.

**HCS** Hundred Call Seconds. One hundred seconds of telephone conversation. See CCS.

**HEADBAND-STYLE HEADSET** The type of headset that uses a plastic, metal or combination band to hold the earpiece next to your hear. If fastens to the head using the same principal as a pair of earmuffs.

**HEADSET** A telephone device that replaces the handset (or receiver). All headsets consist of an earpiece and a microphone — but these elements can be arranged in a variety of ways. Now made of plastic, headsets are light and comfortable. They are so light and comfortable in fact, that executives, consultants and stockbrokers now commonly wear headsets, in addition to the traditional headset users: switchboard attendants, telemarketers, customer service reps and reservations agents. See MONAURAL, BINAURAL and STARSET.

**HEADSET JACK** A place on a phone or console into which you can plug a headset. There are two kinds, the two-pronged jack found on attendant consoles and older ACD station sets, and the RJ-9 modular jack found on electronic telephones.

**HELP DESK** A department or organization that technology users can turn to for help with their glitches, bugs, misunderstandings and general confusion. Some people use "help desk" very specifically. They mean an internal department or division that helps only employees of that company with their questions and problems with the technology (usually computers, software and peripherals) issued by that company. We like to draw the circle a little wider to include all technical support centers, whether they serve internal employees, franchisees or external customers. We feel all these operations share goals, techniques and technologies that are so similar, that for most purposes they can be treated as a whole. (Plus, using one term makes it easier to

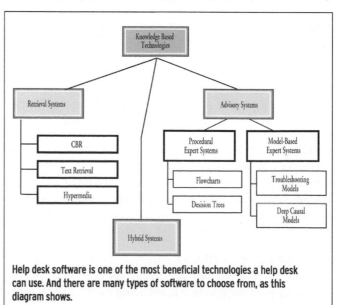

Help desk software is one of the most beneficial technologies a help desk can use. And there are many types of software to choose from, as this diagram shows.

**Which headset is right for your application? Use this guide to find out.**

### Binaural headset

This headband style has two receivers. Sound comes through both ears. This is ideal if you want to block out all background noise.

### Monaural headset

This headband style has only one receiver. It's ideal if you want to keep an ear out to what's going on in the rest of the office.

### Around-the-ear headset

This contour design rests on your ear. It's less obtrusive than the headband styles and won't muss your hair.

### Contoured over the ear style

There are different styles of ear tips that attach to this contoured piece. The ear tip fits in the ear, providing superior sound quality, but some people find them uncomfortable.

### Cordless Headset

A cordless headset weighs more because the battery and transmitter are built in, but it won't keep you chained to you desk. This one's from Hello Direct.

### Amplifier

This is the small box between the telephone and headset. Not all amplifiers work with all phone systems. Be sure to chose one that works with your system. Most models have volume control, a handset and headset switch and a mute button. Some have an on-line indicator, so you know when your agents are on another line.

### Microphone boom

This piece is attached to the ear piece. It hold the mic in front of your mouth. Some are adjustable to let you find the position for the bestsound quality.

describe the increasing popular outsourced help desk.) To specify a center that supports only employees, we use the term "internal help desk." A help desk is the place where product assistance is rendered. It is more than a call center — it can contain libraries, walk-in centers, field technicians and their dispatchers.

Help desk technology can be as simple a paper trouble tickets, index card customer records, selves filled with technical manuals and demonstration models of the supported technologies. But being high-tech people, help desks usually do better than that. Help desk software comes in many flavors: from customer service software; to problem tracking and escalation software; to many different types of sophisticated information databases and problem-solving software. Some of these information databases and problem-solving software packages use cutting-edge artificial intelligence and multimedia technologies to make help desk staffers' jobs easier.

Most help desks share two techniques for solving their customers' problems. The first is problem escalation. Calls come in to the lowest-level help desk agent, who has the least experience and possibly the least training. These agents are generalists. In some help desks they merely assign the problem to an appropriate expert, but many help desks they take a crack at solving the problem using their own knowledge, the solutions provided in a database or manual, and possibly a scripted problem-solving software, which walks them through the questions to ask the customer. If the problem can't be solved by the first-level staffer, it is escalated to an expert in that problem or technology. Large centers may have these experts standing by. Others, especially internal help desks, may refer the caller to the vendor at this point. In some vendor technical support centers the problem winds up, ultimately at the highest source — the engineer or programmer who created the product in the first place. (In other cases there is a team of problem-solving engineers or programmers who make up the highest level. They tear the product apart until the problem is solved.)

The second technique is the collection of a database of solutions. There are various ways of amassing and accessing these databases (most computer-assisted), but the technique remains the same. Known solutions and their symptoms are entered in the database to aid future problem solving. The more solutions that already exist, the easier the help desk agent's job is. Also see KNOWLEDGE BASE, INFERENCE ENGINE, EXPERT SYSTEM.

**HIDDEN MARKOV METHOD** This type of speech recognition compares the spoken input with a model created by taking samples from 500 to 1,000 speakers. The nice thing about this popular method is that well-tested vocabularies already exist. It is used for both discrete and continuous speech recognition. It sounds like the name for a magic trick, and in some ways it is, but it's a modern, high-tech and well-proven magic trick.

**HISTORICAL DATA** Data from a previous logging period.

**HISTORICAL REPORTS** Information about what went on in a call center's operation in the past. This includes call data (duration, efficiency, and agent performance) as well as customer data, and anything derived from the two (like revenue per customer, for example). See REAL-TIME DATA.

**HIT RATE** The percentage of matches a lookup service makes when trying to find phone numbers (or other information) for lists of names and/or addresses.

**HOLD** To temporarily leave a phone call without disconnecting it. You can return to the call at any time, sometimes from other extensions. While the terms "on hold" and "in queue" are in many ways interchangeable, putting a call on hold infers a passive state. It's already routed and waiting to be activated again. When a call is in an ACD queue, many things are happening to route the call to the next available agent. The call is very much in play, active, and waiting, merely to be connected to its final destination.

**HOLD RECALL** A telephone system feature which reminds you periodically that you've put someone on hold.

**HOLDING TANK** A queue in which a call is held until it can either use its assigned route or overflow into the next available route.

**HOLDING TIME** The total time from the instant you pick up the handset, to dialing a call, to waiting for it to answer, to speaking on the phone, to hanging up and replacing the handset in its cradle. You are never billed for holding time. You are always billed for conversation time which is shorter than holding time. But holding time is an important figure to know when you're trying to determine how many circuits you need. For you will need sufficient circuits to take care of dialing, etc. — even though you're not being billed for that time.

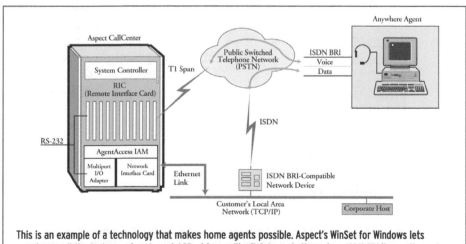

This is an example of a technology that makes home agents possible. Aspect's WinSet for Windows lets agents use all the features of an Aspect ACD at home. The link is made through an ISDN BRI line or through two analog telephone lines

Holding time for inbound calls (the total amount of time they spend on your phone trunks, even when they are not talking to anybody) is important in planning your inbound circuits, especially if you are running a large call center that answers toll-free numbers.

**HOLD TIME BEFORE DISCONNECT** The length of time callers will typically wait on hold before they hang up in disgust. Most ACDs will keep track of this statistic. Hold time before disconnect can be lengthened through the use of music or messages on hold. See AVERAGE WAIT TIME.

**HOLIDAY FACTOR** A historical factor associated with a specific date and multiplied by the forecast call volume for that data in order to take into account an expected increase or decrease in the call volume. For example, if on a given day only half the usual number of calls occur for that day of the week and that time of year, the holiday factor for that date would be .5.

**HOOKSWITCH** Also called SWITCHHOOK or FLASH HOOK. The place on your telephone instrument where you lay your handset. A hookswitch was originally an electrical on/off "switch" connected to the "hook" on which the handset (or receiver) was hung when the telephone was not in use. The hookswitch is now the little plunger at the top of most telephones which is pushed down when the handset is resting in its cradle (on-hook). When the handset is raised, the plunger pops up and the phone goes off-hook. Momentarily depressing the hookswitch (up to 0.8 of a second) can signal various services such as calling the attendant, conferencing or transferring calls.

**HOME AGENT** An agent stationed at a remote site (which could be a satellite center, or the person's home) who is fully connected to the call center's switch. The agent is fully integrated: she shows up on reports and accepts calls exactly the same way as if she was physically in the center. As far as the switch is concerned, the con-

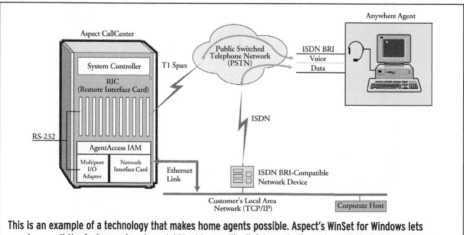

This is an example of a technology that makes home agents possible. Aspect's WinSet for Windows lets agents use all the features of an Aspect ACD at home. The link is made through an ISDN BRI line or through two analog telephone lines

nection is transparent. Also called a teleworker, telecommuter, a virtual agent, or simply an agent-at- home.

**HOST** Usually refers to a computer. In an all-computer setup, it's the computer that drives the terminals. In a computer/telephone setup it's the computer that does the processing that directs other devices in the network (for example, it routes calls using a database) or has the information that can be accessed by other devices.

**HOT CUT** Also known as a flash cut. A hot cut is the instant swap from an old telephone system to a new one. The new system is added to the circuit as soon as the old one is removed. This method of cutover is fast, but very risky. The more conservative method is the "parallel cut," where the two systems run side by side for a month or so.

**HOT SITE** An alternative location used for disaster recovery purposes. What makes the site "hot" and not "standby" or merely "alternative" is the fact that it is maintained so it is always ready to go at an instant's notice. For example, a data storage hot site would have a reasonably close version of all your current data available for backup downloads or to run your systems from if you have a major crash. A telecom hot site would be programmed with all your toll-free numbers and any other needed info. A complete call center hot site would be ready for agents to walk in and start making calls. (Or it may already be staffed with agents, in the case of a service bureau. A service bureau hot site would be ready to roll with all your current data and applications instantly, or almost instantly.)

**HOT-SWAPPABLE** When a component of piece of hardware is hot-swappable, it means that you can unplug that component and replace it while the unit is still running. Things that can be hot-swapped include disk drives, power supplies, basically anything the system can live without.

**HOTLINE** A list segment that contains the most recent buyers. The theory is someone who has just bought something by mail or phone is likely to do it again. The list should specify how long ago its members last bought something. Common lists are "one month hotline," "six month hotline" and "one year hotline." The conventional wisdom is the more recently the people on the list last bought something, the hotter they are.

Not to be confused with the term for a dedicated line or dedicated transmission link between two locations, which was never to our knowledge used by telecom or call center people. These days "hotline" is more often used for a telephone number dedicated to calls of some urgency, for example, a drug abuse hotline. Not a technical term, but of some importance since calls to these numbers are almost always answered by a call center of some kind.

**HOUSE LIST** Your most important list. It is the list your own company has accumulated of its customers. It may include only current customers, or there may be a section for "expires," or past customers. Any list generated in-house is more valuable than a list you can buy. Not only is it free, but it includes your most likely prospects. It's hard to believe, but not all companies keep a house list. Other companies let their house list get out dated or let the last person who understood the database management system leave before the information is extracted into usable form.

Clean your house list often. Use it often. Make sure it stays in good shape. It is a valuable resource for any company that does business by telephone.

**HUNDRED CALL SECONDS** Known by the initials CCS where C is the Roman numeral for "hundred." One CCS is 36 times the traffic expressed in Erlangs. See CCS.

**HUNT** Refers to the progress of a call reaching a group of lines. The call will try the first line of the group. If that line is busy, it will try the second line, then it will hunt to the third, etc. See also HUNT GROUP.

**HUNT GROUP** A series of telephone lines organized in such a way that if the first line is busy then the next line is hunted and so on until a free line is found. Often this arrangement is used on a group of incoming lines. Hunt groups may start with one trunk and hunt downwards. They may start randomly and hunt in clockwise circles. They may start randomly and hunt in counter-clockwise circles.

**HYBRID TELEPHONE SYSTEM** A telephone system that has some of the features of a PBX and some of the features of a key system. As a general (but not strict) rule, a hybrid is a smaller telephone system with a least cost routing (LCR) feature. It may also have some other advanced features that are usually found only in PBXs. Some manufacturers strive for hybrid status for their small PBXs because some (OK, maybe one) local telephone companies charge more for "PBX lines" than they do for key or hybrid system lines.

**HYPERTEXT** Also called hypermedia, especially when pictures are involved. Hypertext lets you create your own path through written, visual or audio information stored on a computer. Certain aspects of the file will be highlighted, indicating there is more information on the subject. When you select a highlighted item, the system leads you to, more information on the selected word, schematic for the axle you clicked on or a picture of the word you selected. It's a popular and useful way to organize technical information for help desk personnel. These days, most people are familiar with it from Internet applications, which let you jump to another page or even another site by clicking on the underlined text.

**ICP** Intelligent Call Processing. The ability of the latest ACDs to intelligently route calls based on several pieces of information, including: information provided by the caller, a database on callers and system parameters within the ACD such as call volumes within agent groups, and number of agents available.

**IDLE** Not busy. Sitting around waiting to handle a call. Used to describe a telephone or telephone agent in a call center.

**IN-BAND SIGNALLING** A method of controlling information in a telecommunications network by using tones or other signals carried within the same band or channel as the information being carried. For example, in a telephone call, tones can be used to control the transmission, receipt and disconnection of the call. If you have ever had a voice mail system cut you off while you were leaving a message (after about three seconds), then you have experienced one of the drawbacks of in-band signaling. Your voice imitated the tone used by the voice mail system to signal "disconnect."

**INBOUND** In this dictionary, a term used to describe calls. Inbound calls are made by other people to your company. From your point of view (or the point of view of your company), they are calls arriving, or coming in, hence "inbound."

**INBOUND CAMPAIGN** A project that receives calls on a designated group of trunks and creates a campaign database from the data entered by the agents.

**INBOUND ROUTING** In the fax world, this refers to the problem of receiving faxes on a fax server connected to a LAN. It's easy to send faxes out from your PC through the server, but when one comes in from outside, the server doesn't know which PC on the LAN to route it to. Often, it has to be printed and manually distributed. Systems equipped with Direct Inward Dialing lines, though, route the fax by assigning each user an extension which is punched in on the sender's fax machine. It also refers to the distribution of calls received by a call center. See DIRECT INWARD DIALING.

**INBOUND WATS** Once this was telecom-speak for 800 or toll free service. The term has fallen out of use except for some old- timers, including former AT&T employees. With inbound WATS service, your callers dial a number with a special area code (800, 888 or 877) and the call is charged to your business, instead of to your callers. Ever wonder why some companies have two 800 numbers, one for instate and one for out-of-state? (We don't think you'll see this with 888 or 877 numbers.) Or why they don't have a toll-free number for instate calls at all? It's because in the old days WATS was strictly a long distance service that used bands of states — and those bands didn't include your home state, you had to arrange that separately. Now things have changed and you can get one phone number for your home state and all the other states too. See TOLL FREE SERVICE.

**INCUMBENT CARRIER** A term used to refer to a national carrier that formerly had a monopoly, but now has competition. It is expected that the carrier will have some advantage in a competitive marketplace since everyone in that country is already familiar with its services. In the United States in 1984, AT&T was the incumbent carrier. Now that telecom competition is being introduced to the rest of the world there are lots of incumbent carriers all over the place. See NATIONAL CARRIER.

**INFERENCE ENGINE** The Artificial Intelligence heart of a knowledge base system. The inference engine is the technology which directs the reasoning process. The inference engine contains the general problem-solving knowledge such as how to interact with the user and how to make the best use of the domain information.

**INFORMATION APPLIANCE** A colloquial term that doesn't actually mean anything in a call center context. Windows Sources defines it this way: " A type of future home or office device that can transmit to or plug into common public or private networks. Envisioned is a 'digital highway,' like telephone and electrical power networks." We have, however, seen it used to describe the modern ACD by people trying to make the point that the ACD is not a piece of isolated dumb hardware, but rather an open, network-spanning piece of futuristic technology.

**INFOMERCIAL** A full-length (usually half-hour) television show created and paid for by the vendor of a product or service for the express purpose of increasing sales. Sometimes a vendor will promote several products on the same show or several vendors will join together to create an infomercial promoting many products. Often these shows ask viewers to call a toll-free number to order the product(s) being promoted. And that, of course, is where the call center comes in.

**INFORMAL CALL CENTER** A term for a group of people who receive or make a lot of calls, but who are not organized as a call center. An example is a group of stockbrokers. They don't call themselves "agents," they may not wear headsets, they don't use an ACD, but their group has many of the same problems and goals as a formal call.

**INFORMATION PROVIDER** A person or company that creates the programming for any on-line service, from Internet services to pay-per-call 900 service. The idea is for the information provider to make money by providing this service. In the telephone version, the provider uses a 900, 976 or 970 code, which signals the caller that a premium will be charged for the call to cover the cost of the information provided. This information can be the weather, technical support for a computer, or even phone sex. The charges then appear on the caller's telephone bill.

**INQUIRY** When a potential customer calls, writes or comes to your office and asks for information on your product or service.

**INSTRUMENT** A telephone set, such as a proprietary ACD phone, a 2500-set, or a terminal for the hearing-impaired.

**INSTRUMENT SIGN-ON** Another term for AGENT ID.

**INTEGRATION** Linking one system (voice or computer) with another so that each system can take advantages of all or most of the features of the other. For example, if your telephone system is integrated with your voice mail system, the voice mail system can take advantage of the phone system's message waiting feature. It will light the message-waiting light when a message is left for that extension. An integrated ACD-to-computer system will not only pass a call to an agent, but can route the call through instructions in the computer and pass a computer file along with the call. For more on this see CTI.

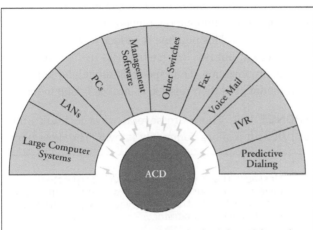

Integration means your ACD does not have to stand alone. Links can be made to a spectrum of call center technologies, from computer systems, to other telephone systems, to voice processing systems. Integration enhances the power of each technology.

When databases are integrated, information you enter or change in one database (say the address field) are also updated in the corresponding field of the integrated database.

**INTEGRATED ACD/PBX** A phone system with enhanced call routing features like those found in standalone ACDs. They are particularly attractive for sites with fewer than 50 agent stations, but they can be much larger. They also offer the opportunity for small centers to expand. Several large PBXs with very sophisticated ACD features are used in inbound call centers as frequently as stand- alone ACDs are. One such PBX is Lucent's Definity G3.

**INTEGRATED SERVICES DIGITAL NETWORK** See ISDN.

**INTELLIGENT OVERFLOW** Software from Aspect Communications that makes intelligent decisions about whether or not to overflow calls based on the traffic load of each agent.

**INTER-APPLICATION AUTOMATION** A term, first articulated by AnswerSoft (a CTI company now owned by Davox) for the way in which well-developed computer telephony middleware products sit between the hardware, the database, and the call center applications they enable. These systems float between layers of technology like the oil in your car's engine.

High-performance switches for call centers are now so open that it's possible to mix and match add-on software for virtually any vertical or horizontal feature preference without sacrificing switch performance. Middleware's most recent role is to stand between the front office call center apps and the back-end databases, in coordination with the switch, to integrate all the current systems into one common environment. This is what is meant by "inter-application" automation: functioning as the data conduit between systems by different vendors. This pipeline is, in fact, a way of connecting not just varying pieces of technology, but two diverse business realms: the inside of the call center and the rest of the organization. They speak different languages and work with different kinds of information, but their mission is intricately tied together.

**INTERACTIVE VOICE RESPONSE** IVR. There are several ways to think about interactive voice response, or IVR as it is more commonly known. The most popu-

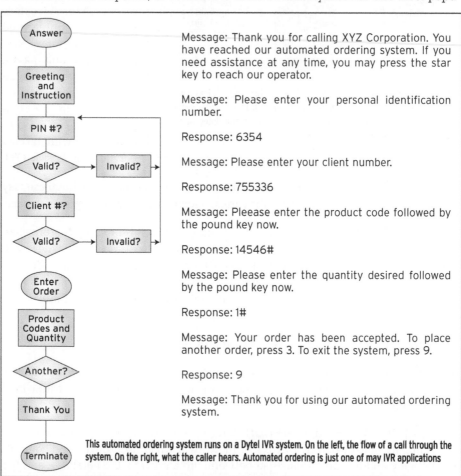

This automated ordering system runs on a Dytel IVR system. On the left, the flow of a call through the system. On the right, what the caller hears. Automated ordering is just one of may IVR applications

lar among IVR gurus is "a telephone interface to a computer system." That is, a system that is stuck on the front end of a computer system that lets you enter information from that system either through a telephone keypad or the spoken word. You receive information through the system through a recorded (and digitized) voice or a synthesized voice. (In some cases you may receive information through fax, or even information on a special screen attached to your telephone.)

Whatever you can do with a computer, you can do with IVR, although there are some limitations. For example, for input, numerical digits are much more simple than any word or letter because of the easy of using the telephone keypad that way. Voice recognition, which lets you speak your input, is making great strides, but still has a limited vocabulary. Output is less limited, although text-to-speech can sometimes sound clumsy and anything graphic must be sent by fax or other special method.

But the benefits are vast. The telephone is familiar to everyone. It already has a world-wide network. Accessing information by telephone lets anyone interact with the computer from anywhere in the world. It also cuts down on the need for agents — especially when repetitive questions and answers are involved. Not only do you save on personnel costs, but you are more likely to keep the agents you like, because their job is less boring.

The classic IVR application takes an existing database (for example, bank records or a freight company's package-tracking system) and makes it available by phone (or other media, such as fax, e-mail, or DSVD — Digital Simultaneous Voice and Data). IVR gives access to and takes in information, performs record-keeping, and makes sales, 24 hours a day — supplementing or standing in for human personnel. From "bank by phone" to "find my package," to "sell me an airline ticket," to "validate my new credit card," IVR is already on the job.

Used as a front-end for an ACD, an IVR system can ask questions (such as, "what's your product serial code?") that help routing and enable more intelligent and informed call processing (by people or automatic systems). IVR far supersedes more rudimentary technologies (such as Caller ID) in such applications. Used in place of traditional on-hold programming, IVR can add interactive value to what would otherwise be wait-time. The latest technology lets your callers play with IVR while they wait for a live agent, while maintaining their place in the queue. Or choose to first use the IVR system, then if they need human assistance, a record of their IVR session appears on the human agent's computer screen, allowing the call to proceed from where it left off.

**INTERBREAK INTERVAL** A scheduling assumption specifying the minimum amount of time that must elapse between the end of one break and the beginning of another.

**INTERCEPT RECORDING** The recording you hear when "your call can not be completed as dialed..." or "The number you have dialed is not in service" or several other

**97**

reasons. The phone company intercepts that call and sends it somewhere. Intercept Recording is a recording explaining why your call did not reach its destination.

**INTERCOM** In the most general terms, this refers to any communication between telephones in the same office or company. You can pick up and dial another extension or you may have a separate system parallel to the office telephone system that lets the boss "buzz" for her secretary or lets the secretary announce a call even if the boss is on the phone.

**INTEREXCHANGE CARRIER (IXC)** A telephone company that provides service from one LATA (Local Access and Transport Area) to another, but not within any one LATA. Essentially, a long distance company. AT&T, MCI Worldcom, and Sprint are all IXCs.

**INTERFLOW** The ability to establish a connection to a second ACD and overflow a call from one ACD to the other. This provides a greater level of service to the caller. See also LOOK AHEAD INTERFLOW.

**INTERLATA CALL** A call that is placed within one LATA (Local Access Transport Area) and received in a different LATA. In other words, a long distance call, not a local call. (Although in some cases, the call does not travel a very long distance.)

**INTERMITTENT PROBLEMS** Intermittent problems are issues or bugs that come to light only after systems have been running for some time, or certain infrequently performed sequences of events are performed. Often many thousands of calls need to be put through before they are discovered. And bugs may only be seen occasionally, perhaps one of every 100 times something is done. Intermittent problems are among the hardest to find and duplicate.

**INTERNAL HELP DESK** A customer support call center whose "customers" are actually employees of the company running the center. Internal help desks differ from other centers in several respects. They are often concerned with facilities management and asset tracking, and so their software often includes those modules. They are also often less well outfitted than external centers; companies have in the past tended to give internal centers short shrift when it came to resources, though there are indications this is changing.

**INTERNET PHONE** Software that sits on a PC and uses an Internet connection to carry a voice phone call. The hardware belongs to the PC: speakers and a microphone. Both parties to the "call" have to be using the same or compatible software (although technology and applications are being developed and piloted that allow a call to begin on the Internet, with someone using an Internet phone, and terminate out in the "real" phone network. See FAX-OVER-IP.

**INTERNET TELEPHONY** Also known as IP telephony, voice over 'net and Web telephony. (And there will probably be more names, both common and trademarked, as it becomes more popular.) Most simply, Internet telephony is sending voice information, or telephone calls, over the Internet, through a PC. There is one huge benefit of using Internet telephony: the cost. However, there are many small drawbacks.

Why is Internet telephony so cheap? Your recurring costs are only the cost of a local telephone call (in some places, nothing), plus whatever you pay your Internet service provider (ISP) each month. You (and the person on the other end) also need a reasonably quick computer, a pretty fast modem, speakers, a sound card, a microphone and some software. (Most PCs have all of that, except the microphone and the software.) Ideally, you should also have a headset, and for your call center agents, that may be the easy part.

The reason why we all don't run out and get the microphone and the software and forget our long distance carriers is the sound quality of Internet telephony. Even though you access the Internet through telephone lines that can comfortably carry your voice, so much other stuff is happening on those lines when you access the Internet that the voice part of the call has to be compressed into a small part of what that line can carry. That means the call doesn't sound anywhere as good. Add to that the fact that the Internet deals with high traffic by dropping packets (not a problem with data), plus the speakerphone-like distortion caused by the microphone and speakers if you don't use a headset, and you get a telephone call that is difficult to understand, filled with static and that has delays and gaps of silence long enough to alter the way you talk on the phone.

In spite of all of this, the sound quality is on par with what a long distance call to some countries sounds like anyway. Some people who regularly call these countries find Internet telephony worth the other hassles, since they really don't notice the difference in sound quality.

And yes, there are other hassles. There is no over-riding Internet telephony standard, and since the voice must be greatly compressed there has to be compatibility between compression techniques. Some Internet telephony software works with just about any other Internet telephony software. Others use proprietary techniques and can only be used between two parties using the same software.

So why might you want to use Internet telephony in your call center? There are a few good reasons. First, if you want to give Web surfers access to a live agent by "phone," sending the voice over the Internet saves the hassles of making the surfer disconnect from the Web to get your callback, the long distance costs of that callback and the general break in the action caused by switching from Web to telephone handset. The lower voice quality is probably a good trade-off for the convenience.

Also, if your call center makes or receives calls from the far-flung places on the globe, especially developing countries or former Iron Curtain countries where your callers are among those with Internet access, Internet telephony might save you money on regular long distance costs. We can't imagine there are many call centers that meet all those criteria, so for now, think of Internet telephony as a good way to talk to your Web surfing customers, and keep your eyes open for future developments.

**INTERNETWORK** Any two or more networks connected by a router. Also, another, more formal name for the Internet. The Internet is a collection of computer networks used by government, industry and educational institutions.

**INTRADAY DISTRIBUTION** A historical pattern consisting of factors for each day of the week that define the typical distribution of call arrival or average handle time throughout that day. Each factor measures how far call volume or average handle time in that half hour or quarter hour deviates from the average half-hourly or quarter-hourly figure for that day. This information enables the program to forecast call volumes in segments smaller than a day and staffing requirements.

**INTRAFLOW** The ability of an ACD to select a second or subsequent group of agents to backup the primary agent group. This lets the caller be served more efficiently and less expensively.

**INTRANET** Your company's internal networks, especially your computer networks. In some cases your intranet is just your LAN. In other cases you have lots of LANs, MANs, WANs and perhaps some private voice networks. Then maybe there is a reason for referring to an "intranet," but most of the time it is used by someone who means, "LANs" but wants to be cool by comparing those LANs to the Internet. The two are, in fact, quite similar.

**IN-WATS** See INBOUND WATS.

**IP** Internet Protocol or Internetworking Protocol. A UNIX-based set of rules that governs the Internet and everything that interacts with it. Any call center software or hardware that operates over the Internet will use this protocol in some way. Not to be confused with the IP that means "Information Provider." See INFORMATION PROVIDER.

**ISDN** Integrated Services Digital Network. A collection of telecommunications transmission standards and services with a goal to provide a single international standard for voice, data and signaling (so one network can be used for all three purposes, instead of having three networks); make all circuits end-to-end digital; use out-of-band signaling; and bring a significant amount of bandwidth to the desktop.

What can ISDN do for your call center? Here are some examples:

Selective Call Screening — Lets you know who is calling before one of your agents picks up the phone. You can answer calls from high- priority customers first. A similar feature tells you who is on the phone even if the line is busy, letting you further prioritize calls in the center.

Shared Screen — Switched data services provided via ISDN let two people in remote locations, both equipped with a computer terminal, to view the same information on their screens and discuss its contents while making changes — all over one telephone line. Great for technical support centers. Have your techs confer with experts all over the country, or simply take a look at the customer's problem.

Network Access — Lets your agents easily access one of your company's databases, even if they don't use it often enough to be on "the network." Think of an agent who only very rarely needs to check a customers credit records before entering a very large order.

Less Down Time/Cost Savings Moves — When a company moves an employee within an office, there can be hours or days of lost production while a computer terminal and phone set are being installed. In some cases, the terminal is connected to a network via coaxial cable. ISDN virtually eliminates down time, as well as the need for coaxial cable.

A key component of ISDN is CCITT Signaling System 7. This is an international telecommunications standard that does two basic things: First, it removes all phone signaling from the present network onto a separate packet switched data network, providing more bandwidth. Second, it broadens the information that is generated by a call, or call attempt. This information — like the phone number of the person who's calling — will significantly broaden the number of useful new services the ISDN telephone network of tomorrow will be able to deliver.

ISDN comes in several flavors (or services). They are:

1. The 2B+D "S" interface (also called the "T" interface). The 2B+D is called the Basic Rate Interface (BRI). The "S" interface uses four unshielded normal telephone wires (two twisted wire pairs) to deliver two "Bearer" 64,000 bits per second channels and one "data" signaling channel of 16,000 bits per second. An S-interfaced phone can be located up to one kilometer from the central office switch driving it.

2. The 2B+D "U" interface. This "U" interface delivers the same two 64 kbps bearer channels and one 16 kbps data channel, except that it uses 2-wires (one pair) and can work at 5 to 10 kilometers from the central office switch driving it. The "U" interface is the most common ISDN interface.

3. The 23B+D or 30B+D. This is called the Primary Rate Interface (PRI). At 23B+D, it is 1.544 megabits per second. At 30B+D, it is 2.048 megabits per second. The first, 23B+D is the standard T-1 line in the U.S. which operates on two pairs. The second 30B+D is the standard T-1 line in Europe, which also operates on two pairs.

4. A standard single line analog phone. A 2500 or a 500 set.

ISDN has been a long time in coming. Some of the frustrations of its unkept promises are reflected in the "other" definitions for ISDN: I Still Don't Know; It Still Does Nothing; Improvements Subscribers Don't Need; I'm Spending Dollars Now.

**ISG** Incoming Service Grouping. A fancy name for hunting or rollover. You receive many incoming calls. You don't want to miss a call, so you set your phone lines up to roll over, also called hunt, also called ISG in telephonese. You order five lines in hunt. The calls come into the first. If the first one is busy, the second rings. If it's busy, the third rings. If they're all busy, then the caller receives a busy.

There are two types of hunting: sequential and circular. Sequential hunting starts at the number dialed and ends at the last number in the group. Circular hunting looks at all the lines in the hunting group, regardless of the starting point. Circular hunting, according to our understanding, circles only once (though your phone company may be able to program it to circle a couple of times). The differences between sequential and circular are subtle. Circular seems to work better for large groups of numbers.

You don't need consecutive phone numbers to do rollovers. Nowadays you can roll lines forwards, backwards and jump around. Rollovers are now done with software. (They used to be done electro-mechanically.)

**ISP** Internet Service Provider. A company or other organization that provides you with access to the Internet. You may be a corporation or, well, just yourself. An ISP does a number of things for you, including storing World Wide Web home pages, handling your e-mail, and administrative details such as tracking and billing your connections, and securing user names and passwords. The most important thing an ISP does is act as a server for your interaction with the Internet.

**IT** Information Technology. Yet another term for your company's computer department. See MIS.

**ITU-T** International Telecommunications Union. (We haven't been able to find out what the second "T" stands for.) This is a United Nations-based organization that now creates the telecommunications standards that everyone listens to. It has taken over that role from the CCITT. The ITU started out regulating satellites, and its prominence in all international telecommunications regulations now is a testament to the importance of satellites in all aspects of telecommunications. See CCITT.

**IVR** See INTERACTIVE VOICE RESPONSE.

**IXC** IntereXchange Carrier. A fancy way of saying "long distance carrier."

**JUNK FAX** The fax equivalent of junk mail. It's material sent to someone's fax machine that he or she never asked for, and worse, don't want. It should be every direct marketer's goal to avoid junk mail, junk fax and junk phone calls. These customer contacts do not make you friends — but they still cost you money, even if it's not a lot of money compared to how much you will make from a potential sale.

Junk fax is probably the worst of the three and is most likely to create an enemy for life. The recipient of a junk fax is forced to pay for the (very expensive) paper the message is printed on, and it ties up their machine. Always ask before you send promotional faxes, and try to send them at night, when the machine is likely to be free.

True junk faxes are illegal by federal law. This law says you can only send faxes to people you have already done business with and must stop sending the faxes if asked.

**K** Short for the metric prefix "kilo," meaning thousand. If you make 50K a year, you make $50,000. In computerese, K is equal to 1024. This is because computers work in base 2, and 1024 is a power of 2. One kilobyte is equal to 1024 bytes. See MEGABYTE.

K is also used by some telecom types to refer to a K-style handset. See K-STYLE HANDSET

**K-STYLE HANDSET** A telephone handset either has a square mouthpiece and earpiece or round ones. Newer phones, as a rule, have the square handsets, called K-style. (The round kind is called G-style.) If your headset dealer asks you if you have an electronic or carbon-based microphone in your telephone, your first clue to answering this question is the shape of your telephone mouthpiece. If it is square, chances are excellent the phone has an electronic microphone. If it is round, it's carbon-based. Be warned, while the round mouthpieces screw off, the square kind don't come off at all. See G-STYLE HEADSET.

**KEY** The physical button on a telephone set. What normal people call a "switch," as in on and off switch, telephone people call a "key." A collection of telephone keys is found on a keypad. See KEYPAD.

**KEY PERFORMANCE INDICATOR** KPI. Those statistics (and other factors) that are determined to be most important to figuring out how good a job one of your agents is doing. The same or other KPIs can be used to figure out how well a group of agents or the whole call center is doing. For example, you may decide that a friendly tone of voice, the number of calls answered and the number of complaints or comments about an agent are the most important factors in how he or she does his job. Those three things would be your KPIs. Or you might decide the depth of knowledge, number cases closed and average talk time are the KPIs for your center.

KPIs can come from ACD statistics, statistics collected by your applications software, or a supervisor's monitoring or work review.

**KEY SERVICE UNIT** See KSU.

**KEY SYSTEM** A type of telephone system. It's usually very small with just a few lines and extensions. To place a call outside the premises with a key system you must choose that line by pressing a button. You may have a button for the local telephone trunk, another for a long distance trunk. If you have more than one long distance service, it's up to you to chose between them.

For incoming calls, the phone will ring and a button will flash. Hit the button and you've got the call. You can pick up any extension that appears on your telephone just by pressing that button.

In the old days, key systems were for the smallest business and for larger companies who needed some departmental functions behind Centrex or a PBX. With the advent of Nortel's open Norstar key system, a few vendors have jumped in and created sophisticated ACDs or integrated inbound/outbound system using the Norstar and a PC, making the key system a viable call center technology.

**KEYPAD** The dialing mechanism on a modern telephone. It is always arranged configuration of three columns and four rows of push- button keys. These keys are labeled with the numbers zero through nine, an asterisk symbol and a pound or numeral symbol. In North America these keys also bear the letters of the alphabet, starting with A, B and C on the two key. Be warned, telephones in other parts of the world either do not have letters or arrange the letters differently. Note that the number in the upper left hand corner of your telephone keypad is a one. On the keypad on your calculator, adding machine or computer, it's a seven. The numbering sequence on telephone keypads is completely different from other keypads.

**KILL MESSAGE** A recorded message played at the beginning of a call to a 900 (or other pay-per-call) number that warns the caller of the charges and gives him an option to hang up before it starts.

**KILLER APP** The application that makes a technology irresistible to the market. For example, no one had a video tape player until people realized they could see rented X-rated movies with little fuss. The market moved on quickly from there. Many call center technologies are still looking for their killer apps.

**KNOWLEDGE BASE** A help desk or technical support term. It's a collection of knowledge about a particular subject, usually in question and answer format, or as a series of if-then statements. "If the paper jams in the printer, then unplug it." The system uses artificial intelligence (AI) to mimic human problem solving. It applies the rules stored in the knowledge base and the facts supplied to the system to solve a particular business problem. See EXPERT SYSTEM.

**KNOWLEDGE WORKER** A person who analyzes, enhances or otherwise manipulates data — usually on a computer. Knowledge workers commonly include computer programmers, consultants of all kinds, writers, editors, engineers and (drum roll, please) just about every kind of call center agent. You can easily see how a help desk agent fits into this. He or she accesses lots of information on technology from various sources, analyzes that information in light of the caller's problem and creates a solution — while leaving an information record of the whole process. But you may not realize how much simple call center tasks like order taking and customer service are really information-intensive transactions, facilitated by the agent. They are.

**KPI** See KEY PERFORMANCE INDICATOR

**KSU** Key Service Unit. A small metal box, often found mounted on the wall in the "phone room" that provides the guts and the brains (such as they are) of a business key telephone system. This is where the actual switching takes place. It's the system's central processing unit. A few basic things every call center person should know about a KSU are: it needs to be well-ventilated, it should be near a power outlet, and it should have it's very own power circuit, which is not shared with photocopy machines or air conditioners.

**LAI** See LOOK AHEAD INTERFLOW

**LAN** Local Area Network. A system connecting a set of computers and peripherals over short distances. It allows users at multiple computers to use the same files and share printers. A LAN will typically link devices in a single building, but it can stretch as far as about 10 kilometers. Larger than that and it is called a MAN (Metropolitan Area Network) or WAN (Wide Area Network), which have different properties.

**LAST IN FIRST OUT** LIFO. The last phone call arriving is the first call to leave — to be processed, to be saved, whatever. The term LIFO comes from accounting.

**LATA** Local Access and Transport Area. One of 161 local geographical areas in the US within which a local telephone company may offer telecommunications services — local or long distance. Whether long distance companies such as MCI or Sprint, can carry intraLATA calls varies by the rules of each state.

**LCD** Liquid Crystal Display. A screen-display technology that uses a crystal layer suspended in a liquid under the influence of a low voltage so it reflects ambient light and displays alphanumeric characters. Benefit: it requires very little power. Drawback: it can be hard to read from certain angles. It is also difficult to read under certain light conditions unless it's back-lit or illuminated in some other way. But doing this requires more power...

On a telephone, this benefit means the display can be line powered — that is, powered by the one or two pairs coming from the ACD or other telephone system. Such LCD displays on electronic phones can perform many functions. On ACD station sets they have been used to display queue statistics (how many calls are in queue, longest call waiting) and other call information, such as the company associated with the number dialed by the caller (helpful in service bureaus and centers that handle multiple product lines).

**LDT** Longest delay time. An ACD statistic. The longest time any caller waited for an agent — whether the caller abandoned the call or was served by an agent.

**LEAD AGENT** The first agent in an ACD group. See also AUTOMATIC CALL DISTRIBUTOR.

**LEAH** The brightest star that shines on earth, according to her mother.

**LEAST-COST ROUTING (LCR)** A feature of a telephone system that automatically connects an outgoing telephone call with the telephone service that will cost the least to that location at that time of day. Depending on how it's programmed it will drop down to the second most efficient service if the first is not available or it will give the caller a busy signal.

Least cost routing eliminates guessing and stupid mistakes (using a carrier that costs 10 times as much for that call) by employees, but it's got to be programmed correctly for the services you have. LCR is the main difference between PBXs and key systems.

**LEC** Local Exchange Carrier. A local phone company, which can be either a Bell Operating Company (BOC) or an independent (for example, GTE) which provide local transmission services. Prior to divestiture, the LECs were called telephone companies or telcos.

**LED** Light Emitting Diode. A semiconductor diode that emits light when a current is passed through it. It is used in a lot of applications, from data transmission to readouts on digital equipment (like watches and calculators). LED technology is also used as a lower cost substitute for lasers in page printers. LEDs use less power than normal incandescent light bulbs, but more power than LCDs (Liquid Crystal Displays).

**LEGACY SYSTEM** A mainframe or minicomputer information system that has been in place for far longer than everyone would like. The computer (or telephone system) left over from a previous manager's reign, which you now have to deal with.

**LEGAL HOLIDAY** Any holiday for which special wages (time and a half, double time) are paid to agents who work on that day.

**LIBERATION** A formerly Nortel line of headsets designed for use with Meridian telephones and attendant consoles. The Liberation product line includes monaural and binaural styles. It's now been taken over by GN Netcom.

**LIFE CYCLE** A term used in STRUCTURED WIRING (the coordination of wiring plans within a call center). How long the cable is physically anticipated to be in place. For example, if a customer intends to be in a large office for 10 years, fiber installation may be considered.

**LIFO** See LAST IN FIRST OUT.

**LIGHT EMITTING DIODE** See LED.

**LIKERT SCALE** A way of phrasing a survey question so that it will provide quantifiable data, yet also provide more detail than a simple yes or no answer. The question could be, "How often does your shipment arrive on time?" The answer: "Always, often, sometimes, never," is a Likert scale. Other common Likert scale answers ask respondents to say how much a statement reflects their point of view, "Strongly agree, somewhat agree, somewhat disagree, strongly disagree." A variation has opposites (like, dislike) at either end, with a series of numbers (0-6) corresponding to the level of like or dislike.

**LINE** The word "line" is confusing. In traditional telecom, a line is an electrical path (two wires) between a phone company central office and a subscriber, usually with an individual phone number that can be used for incoming and outgoing calls. A line, in this definition, is the most common type of loop. A line is also a family of equipment or apparatus designed to provide a variety of styles, a range of sizes, or a choice of service features. As in "product line." The confusion over the word "line" starts with an office phone system. Some people believe a line to be the same animal as a trunk — that is the line coming in from the central office to the ACD or other telephone switch. Other people think a line is an extension, that is the line from the ACD to the phone on the agent's desk.

**LINE-POWERED** A device, like a telephone, that draws the electrical power it needs for use from the phone line it is connected to. The advantage, obviously, is that you can use these devices in places where there is no electrical socket. A small modem, for example, can be designed to use power from the phone line so you can use it at a pay phone or other inconvenient spot. Some modems that call themselves line-powered actually use power from both the line and from the computers they connect to. Headsets are a call center device that can be line powered — but usually only if they are used with a carbon-based phone. The big question then is whether to use a plug-in power pack (clutters the desk, makes a terrible tangle of wires) or a battery power pack (runs out much too quickly for the average call center agent's use).

**LINESHARING** Using the same phone line for different kinds of transmissions, like voice and fax, voice, data modem, and answering machines. There are devices (hardware and software) that can watch a line and let calls go through to the intended recipient. Some work by setting each machine to answer on a different ring number. Others detect the tone coming from the other end and route the call accordingly.

**LIQUID CRYSTAL DISPLAY** See LCD.

**LIST APPENDING** See TELEPHONE NUMBER APPENDING.

**LIST BROKER** A liaison between the list owner and potential list renter. The broker "works" for the renter by suggesting appropriate lists to the renter and is paid a commission by the list owner.

**LIST BUYER** The person or company who rents, leases or uses a list — even if it's just once. It can also mean someone who actually buys a list, but not usually. See LIST RENTAL.

**LIST CARD** Also called a list data card or sheet. This card includes important information about a list, including: source, quantity of names available, cost per thousand and suggested markets.

**LIST ENHANCEMENT** Adding information (such as phone numbers) to a list to improve its performance or value. List enhancement can be as simple as adding telephone numbers, or as complex as adding a virtual biography of demographic information to each person on the list.

**LIST LOOK-UP** See LOOK-UP SERVICES.

**LIST MANAGER** A person who takes care of list rental details from promotion to collection for the list owner.

**LIST PENETRATION RATE** Refers to rate of successful calls made to numbers on a list through outbound dialing. Here's an example: with manual dialing, a company could buy a list of 100,000 names and reach half of them (50% penetration) by trying each name three times. But with advanced dialers, they have the time to make 30 or 40 attempts at a difficult number, but they achieve 90% penetration. The upshot — they can buy smaller lists and cut costs.

The list enhancement process has been improved through electronic transmission, using modems and electronic bulletin board systems. The result is faster turn around time.

**LIST RENTAL** An agreement between the list owner and a mailer to use a list — usually for one time only and at a certain rate per thousand names. Mailing lists are almost never really "sold." Selling a list (and owning one) implies the right to rent the list out to others.

**LIST SELECTION** Criteria used to flag a part of a list. For example, you may rent a list of fishermen because you are selling a fish finder. The gizmo is very expensive, so you want a selection based on the monetary value of the last purchase. You're only interested in those fishermen who spent more than $300. Thanks to list selection, you now have a smaller, but more responsive group. When buying a list, keep in mind that the more selective it is, the more expensive it is.

**LIST SEQUENCE** The order of names on a list. Most lists are in Zip Code order to make meeting Third Class mail rules easier — but this term can refer to any kind of order in a list. Your call center's list sequence may be by area code, so you can dial calls in a certain time zone during a specific time period.

**LIST SORT** What you do to get your list into a specific list sequence.

**LIST SOURCE** On an in-house list this is the media or method (telemarketing, TV advertisement, direct mail) that generated the names on the list. On a rented list it is the organization that generated the list. If you rent a list of subscribers to Cat Fancy magazine, then that magazine is the list source. If anyone from that list responds to your offer you may want to keep track of the ultimate source of that name for future reference.

**LOAD BALANCING** The practice of splitting communication into two (or more) routes. By balancing the traffic on each route, communication is made faster and more reliable. In telephone systems, you can change phone and trunk terminations to even out traffic on the network. An example: You have call centers in three locations, each served by an ACD joined by tie lines. Instead of having each ACD route only the calls it receives, you spread the traffic over all three ACDs for faster processing. The objective is to have the least number of calls blocked (receive a busy signal) and not wear out agents at one center while the others sit idle.

**LOCAL ACCESS TRANSPORTATION AREA** See LATA.

**LOCAL AREA NETWORK** See LAN.

**LOCAL CALL** Any call within the local service area of the calling phone. Individual local calls may or may not cost money. In many parts of the US, the phone company bills its local service as a "flat" monthly fee. This means you can make as many local calls per month as you wish and not pay extra. Increasingly this luxury is dying and local calls cost money.

**LOCATOR** Another way of saying "dealer locator." A popular interactive voice response application that uses a database lookup (based on telephone number, zip code or other geographic identifier) to tell the caller where the nearest dealer for the product in question is located.

Also used in the secondary telecom equipment business to describe a company that assists both a buyer and seller to quickly find each other. A locator contracts with dealers to provide them with daily lists of potential customers.

**LOG-ON IDENTIFICATION** A code assigned to an agent that allows the ACD to collect statistics about what that agent does once he or she signs into the system.

**LOGGING PERIOD** A period of time (usually 30 minutes in length) used by Rockwell's Galaxy and Spectrum ACDs as the basis for statistical reports.

**LONG DISTANCE** Any telephone call to a location outside the local service area. Also called toll call or trunk call.

**LONGEST AVAILABLE** This is a method of distributing incoming calls to a bunch of people, such as in an ACD. It selects an agent based on the amount of time that each agent has been off the phone. When a call comes in, it is routed to the agent that has been off the phone the longest. A simple and reliable way to distribute calls equitably. In the modern call center, however, much more sophisticated routing schemes are possible. See also TOP DOWN, ROUND ROBIN and SKILLS-BASED ROUTING.

**LONGEST CALL WAITING** A very basic ACD statistic, usually given for each queue, which shows the amount of time the caller who has been on hold the longest has been waiting. This statistic is often found on readerboards posted in the call center, so agents know how fast they should work. Many centers have goals to have no caller wait longer than 30 seconds or five minutes. When this goal is approached or exceeded, alarms (visual or audible) go off.

**LONGEST DELAY TIME** See LDT.

**LOOK AHEAD INTERFLOW** A feature of an inbound switch that lets you route calls across multiple sites. The routing is not done by or in the long distance carrier's network but your own. One of the switches in your network keeps track of the call traffic at all your sites. When a call comes in it is routed to the switch and the center that can best handle that call, according to your previous programming and instructions.

**LOOK-UP SERVICES** Services that take your list and do one, or more, of the following: matches your list of names, addresses and phone numbers to theirs and spits out any differences; takes your name (perhaps someone who has moved) and adds their telephone number or address; takes a name or phone number not in any electronic directory and searches the old fashioned book directories for that name. These services are closely related to list enhancement, and are often provided by the same companies. The nuance seems to be list enhancement leans toward psychographic information and is done by list companies while look-up services stick to the nuts and bolts and are provided by electronic directory companies. More and more call centers are using real-time look-up services to match the telephone number of an incoming call with a name and other information. This feat is called a "data dip."

**LOOPBACK TEST** A loopback is a diagnostic test usually run on a four-wire circuit. The transmitted signal is returned to the sending device after passing through a data communications network so the send and receive signals can be compared. Often the test is run again and again, excluding one piece of equipment after another until the defective one is found.

**LOOP START** LS. You "start" (seize) a phone line or trunk by giving it a supervisory signal. That signal is typically taking your phone off hook. There are two ways you can do that — ground start or loop start. With loop start, you seize a line by bridging through a resistance the tip and ring (both wires) of your telephone line. The Loop Start trunk is the most common type of trunk found in residential installations. The ring lead is connected to -48V and the tip lead is connected to 0V (ground). To initiate a call, you form a "loop" ring through the telephone to the tip. Your central office rings a telephone by sending an AC voltage to the ringer within the telephone.

When the telephone goes off-hook, the DC loop is formed. The central office detects the loop and the fact that it is drawing DC current and stops sending the ringing voltage. In ground start trunks, ground starting is a handshaking routine that is performed by the central office and the PBX prior to making a phone call. The central office and the PBX agree to dedicate a path so incoming and outgoing calls cannot conflict. Here are two questions (and answers) that explain a little more:

How does a PBX check to see if a CO Ground Start trunk has been dedicated?

To see if the trunk has been dedicated, the PBX checks to see if the tip lead is grounded. An undedicated Ground Start Trunk has an open relay between OV (ground) and the tip lead connected to the PBX. If the trunk has been dedicated the CO will close the relay and ground the tip lead.

How does a PBX indicate to the CO that it requires the trunk?

A CO ground start trunk is called by the PBX CO caller circuit. This circuit briefly grounds the ring lead causing DC current to flow. The CO detects the current flow and interprets it as a request for service from the PBX.

**LOST CALL ATTEMPT** A call dialed put not answered at its ultimate destination due to an equipment shortage or failure in the network.

**LOST CALLS CLEARED** Traffic engineering assumption used in Erlang C that calls not satisfied (answered) on the first attempt are held (delayed) in the system until satisfied.

**LOST CALLS HELD** Traffic engineering assumption used in Poisson that calls not satisfied (answered) on the first attempt are held in the phone system for a period not exceeding the average holding time of all calls.

**LOW BATTERY CUTOFF** A power protection term. Refers to automatically shutting off battery power before the batteries discharge beyond safe limits. Without this feature, batteries can be deep discharged, making them useless.

**LUCENT** Lucent itself says, "Lucent Technologies, headquartered at Murray Hill, NJ, designs, builds and delivers a wide range of public and private networks, communications systems and software, data networking systems, and business telephone systems and microelectronics components. Bell Laboratories is the research and development arm of the company."

We say, when AT&T broke up (the second time) it broke into three pieces: long distance services, telecommunications gadgets and computer gadgets. Lucent is the telecommunications gadgets part. It makes what is, at least the last time we checked, the best-selling ACD on the market, the Definity G3. The Definity is not really an ACD you say? (Then you probably work for a competitor.) The Definity G3 may or may not be a "true" ACD, but a whole lot of people use it in their call centers. See AT&T.

**M** Preferred by list people for the abbreviation for one thousand. It comes from the Latin word "mille" (which means "thousand"). Technical people prefer the abbreviation "K" — which doesn't always mean the same thing.

**MAIN DISTRIBUTION FRAME** The point at which outside plant cables terminate, and cross-connections are made to terminal or central office line equipment.

**MAINFRAME SERVER** A mainframe server is a large computer that stores lots of information and manages libraries of information. Employees accessing this information, use "client" computers, usually a PC. Clients are devices and software that request information.

**MAKE-BUSY** You don't want your call center to use a particular circuit, terminal, or termination at this time. To make it unavailable, you make it look busy to the circuits that are trying to connect to it.

**MANUAL GAIN CONTROL** MGC. There are two electronic ways you can control the transmission of an electronic signal — Manual or Automatic Gain Control (AGC). AGC is an electronic circuit in tape recorders, speakerphones and headphones which is used to maintain volume. AGC is not always a brilliant idea since it attempts to produce a constant volume level. This means it will try to equalize all sounds — the volume of your voice and, when you stop talking, the circuit static and/or general room noise which you undoubtedly do not want amplified. When recording something that you want to sound professional (an auto attendant greeting or a message on hold), it is often better to use manual gain control, which lets you turn the volume down and up by hand as needs warrant.

**MANUAL LOOKUP** A telephone number matching technique that enhances lists by finding phone numbers by having people search through phone books or call directory assistance. It is slower than computer-based lookup, but in many cases, a person will find numbers a computer can't. A good approach to finding the most matches for a list is to run it through a computer to find the obvious ones (usually about 50% of a list) and then have a manual service look for the rest.

**MARGINAL NAME** A name that could or could not go on a list says mail. You may leave the name on the list to take advantage of a postal discount. Of course, if you are making telephone calls, this is the name you want to cut to save money.

**MARK** A name for a man who should know by know that when you are really good, you can never get what you deserve in life, or in print. The world's greatest husband and father.

**MASTER FILE** The big file that includes all the information you have on other, smaller files.

**MATCH CODE** A code you use to select records for processing or remove records addresses. You may want to take certain addresses, such as prisons, off your list. You wouldn't want to offer a $5,000 credit limit to someone in prison — although stranger things have happened.

Some companies use a matching algorithm (rather than a code) that doesn't require direct match to select names or addresses from a list.

**MATCHES** When using a large list or database, you will often want to augment it with names, addresses or other information. A match is a correct association of one record with a piece of data in another database, like finding a phone number for a person.

**MAXIMUM QUEUE LENGTH** See MQL.

**MDU** See MESSAGE DISPLAY UNIT.

**MEAN TIME BETWEEN FAILURE** See MTBF.

**MEAN TIME TO ABANDON** See MTA.

**MEANINGLESS DATA** Information gathered from a survey or other market research effort that cannot be used because it is either corrupted or ambiguous. Data can be corrupted by errors or by bad interviewing techniques. Ambiguity can arise from poorly phrased questions, questions that respondents don't understand, or having too small a sample. If you have a very high percentage who respond "Don't Know" or "No Answer," you might have a meaningless survey.

**MEDIA BLENDING** The ability of an ACD to accept and route incoming traffic from a variety of sources, beginning with standard telephone calls, and expanding outward into IVR, other forms of voice response (including speech recognition), video kiosks, faxes, e-mails and interactive Web connections. Needless to say, these are rarely seen all in the same center. Most centers have at most two or three variations on the standard telephony traffic, but even so, coordinating them and routing them to the right person in a timely manner is hellishly complicated.

**MEDIA CONSOLIDATION** The process of bringing customer contact into a call center through a wide variety of channels, including standard telephony, fax, e-mail, Web hits, even video and other esoteric media. This is perhaps the best term; let's hope it replaces clunkers like "multiple access channels, " "media blending," the very horrible "media convergence" and even the one we invented, "customer contact zone" to describe the fact that a call to a call center can come in many forms.

We have to credit Larry Byrd at Quintus for bringing this one to our attention.

**MEGACENTER** An MCI definition. An MCI facility providing concentrated tele-marketing. A megacenter does not handle incoming calls or customer service.

**MEMO** A free form field used to store descriptive text or comments — especially in sales software. The information in a memo field can be of any length and type.

**MERGE** A process that combines information from two or more files (or fields within files) to create a new list. You might draw names from one list, match them with phone numbers from another, and output the result to a file. A related process — "merge/purge" — involves searching the database for names or addresses that should be removed because they are duplicates or wrong addresses.

| | |
|---|---|
| Number of callers per day | |
| Average hold time or time in queue | |
| Daily on-hold total (seconds) | |
| Convert seconds to minutes | divide by 60 |
| Daily on-hold total (minutes) | |
| Average number of business days per month | x 22 |
| Monthly on-hold total (minutes) | |
| Twelve months per year | x 12 |
| Annual on-hold total (minutes) | |

How much time are your callers spending on hold? Find out by using this worksheet, courtesy of On-Hold America.

**MERGE AND PURGE** Putting together two or more lists (merging) and eliminating the duplicate names that result (purging). This process helps eliminate duplicates when you plan to use several lists for a single campaign. It prevents you from calling the same person twice during the same campaign.

**MERLANG** A registered trademark of Pipkins, Inc. (St. Louis, MO), Merlang stands for Modernized Erlang. The company also uses the registered trademark Merlang-M for its multiqueue formula. As the name suggests, Merlang is based on the commonly used Erlang formulas (See ERLANG FORMULA), but adds into the formula busy signals and abandoned calls by taking into account the number of trunks available and an average time callers will wait for an agent until they abandon the call. Pipkins calls this average delay to abandon Mean Time to Abandon, which is admittedly a more precise term.

**MESSAGE DISPLAY UNIT MDU.** Rockwell's term for a standalone device designed to be centrally located among a pool of console positions for the purpose of displaying operating information, such as supervisor messages and gate status data. In Rockwell's case, it is capable of displaying 17 alphanumeric characters at one time and scrolling messages of up to 36 characters in length. In other cases, this is called a readerboard or a display board.

**MESSAGE-ON-HOLD** A recording played to callers while they wait in an ACD queue or while they are put on hold by an agent. A message-on-hold fills a number of functions. It assures callers that their calls have not been disconnected. It entertains the callers while they wait, so they don't hang up. It prepares callers for the upcoming transaction, asking them to have credit cards and order numbers ready. It answers frequently asked questions. (Great for help desks.) It promotes the business, or advertises new products and services.

These days messages-on-hold are almost always played on a digital announcer, which stores the message on a computer chip after it is recorded or downloaded from a cassette tape. Your call center can create its own messages-on-hold or it can hire an outside company which will create the script, provide musical interludes with copyrights already cleared, hire voice talent to record the message for you, and professionally record and mix the final product. You can select one, a few or all of these services from a message-on-hold company.

**METHODOLOGY** A term used in STRUCTURED WIRING (the coordination of wiring plans within a call center). The physical means of getting the wiring system to the user (its distribution path). Examples include modular furniture, surface mounts, fixed wall, recessed wall, raised floor and undercarpet wiring.

**METRICS** A fancy word for "measurements" or "important measurements." Usually used in the phrase "call center metrics," which are those measurements that are vital to figuring out if the call center is doing its job and how well it is doing its job. Service level is the measurement most often included among call center metrics, but these measurements might also include speed of answer, abandons, agent occupancy, sales made or others.

The term metrics is also used for those measurements or statistics that are vital for figuring out if a particular agent is doing a good job. In this case the metrics are drawing a picture — in numbers — of the ideal agent. These statistics usually include agent occupancy, average talk time, number of calls answered plus results of monitoring or call reviews by the agent's supervisor.

In either case, metrics in an attempt to make an objective measure of quality by analyzing statistics.

**MIDDLEWARE** Software that understands the languages of both computers and phone systems and converts messages from one format to another. It enables an application running on a desktop computer to send instructions to a phone system, and vice versa. It sits sandwiched between two other systems, facilitating their communication.

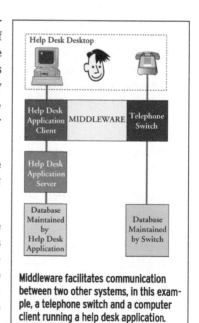

Middleware facilitates communication between two other systems, in this example, a telephone switch and a computer client running a help desk application.

**MIGRATION PATH** Today you are a small company, but someday you will be a large company. Or today you are a big company, but someday you will be huge. What happens when you outgrow your telecommunications or computer equipment?

The manufacturer who is interested in keeping your business as you grow will present you with a migration path. Keep the station sets but add a new central unit. Keep the hardware but upgrade the software. Add new modules and serve more users. Or maybe just buy the next product in the line.

The best migration path is the one that lets you keep all or most of your old stuff and just add more as you need it. A bad migration path will have you junking the whole system and buying from another manufacturer.

MIS Management Information System. A fancy way of saying computer, or sometimes data processing. Your company's MIS department is the department in charge of computers, software and peripherals. A call center manager should have an excellent relationship with the MIS department. Some call centers have one or more MIS people dedicated to the center itself. These people make changes to call center software when needed, and do adds, moves and changes in computer equipment for agents and administrators. Some ACD manufacturers refer to the data cranked out by the switch, and your method of manipulating that data as "MIS." Universities also have MIS departments, but there it stands for "Management Information Science."

**MIS REPORTS** The collated information about calling patterns in a call center generated by an ACD. They will usually include a breakdown of call volume per campaign, per agent group, or even per agent. They can also track data on trunk usage, peak calling times, and the number of times the volume passed the acceptable threshold.

**MODULAR JACK** This is the receptacle for a modular plug. "RJ-11" is the technical name for the common plastic gizmo that can hold up to six wires and allows you connect and disconnect telephone gadgets with the flick of the plastic prong on the top or bottom.

Unplug your phone at home and chances are you'll see just two wires (red and green) housed in the plug. At work you may find two, four or even six wires in there depending on the sophistication of the phone system and what is being carried to your telephone.

If a device advertises it connects with a modular jack (or plug) what they are telling you is that it is easy to install and you probably won't need any extra hardware to plug the thing in.

**MONAURAL** A headset with just one earpiece.

**MONETARY VALUE** Information given about people on a list that is for rent. How much money a direct marketing customer spent on his or her last purchase. Or a price range or the customer's whole buying history in terms of dollars and cents. An average may also be available for the whole list. A very important criteria to consider when renting a list.

**MONITOR** To listen in on a conversation to evaluate the quality of the agent's interaction with the customer. Conversations may be monitored for politeness, clearness of diction, accuracy of information or adherence to a script. In many call centers, monitoring is the most important method for evaluating agent performance.

Monitoring can be performed in several ways. A supervisor can walk up to an agent station and plug an additional headset into the telephone (if there is a jack for this purpose). The agent is well aware of being monitored and the supervisor may be squeezed into a cubicle with the agent. Some systems let the supervisor select an agent for monitoring from a supervisory station set or terminal. There are also systems which allow you to record agent conversations for review later.

Federal legislation has been pending for years that would eliminate monitoring as we know it. (That is, monitoring when the agent is unaware of being monitoring.) This legislation has been brought forth by directory assistance operators and long distance service operators, who have a very powerful union. Getting federal legislators to understand the difference between these operators and the average call center agent has been a huge task for lobbyists for various call center trade groups.

For the time being, it is an excellent idea to have all your agents sign an agreement that shows they understand that having their business telephone calls monitored is part of their job. Make company policies about personal calls during working hours clear. Tell supervisors that they should disconnect from an agent's personal calls immediately, even if that call is in violation of company policy. Review call evaluations with agents promptly. Have an established criteria for call quality that agents are aware of before their first call evaluation.

**MONTHLY FACTORS** A historical pattern consisting of 12 factors, one for each month, that tells the program how much that much that month can be expected to deviate from the average monthly traffic year after year. For example, a monthly factor of .75 means that the month will be 25% slower than average, while a factor of 1.15 means that the month will be 15% busier than average.

**MPEG** Motion Picture Experts Group or Moving Pictures Expert Group. A format for compressing and storing full-motion video very similar to the familiar JPEG format used to store still images, especially for Internet use. There are two versions. MPEG-1 is for computer games and such. MPEG-2 is for broadcast quality video applications such as video conferencing. Important if you are interested in offering your customers information by video, whether that is a video conference with an agent or pre-recorded video information.

**MQL** Maximum queue length. The largest number of calls waiting at one time for an agent group or split.

**MTA** Mean Time to Abandon. A term used by Pipkins, Inc. that means much the same as the more widely used "average delay to abandon" or ADA. The Pipkins term is more exact, giving you the type of average used (the mean). Both terms can be defined as how much time, on average, your callers are willing to wait for the next available agent before they hang up.

**MTBF** Mean Time Between Failures. How long it takes in laboratory tests (on average) for the thing to break down or glitch. Always expressed in a unit of time that makes it seem as though the device will never break down in your lifetime (such as millions of seconds). Quickly convert to a unit of time that has some meaning (days, months, years) before you allow yourself to be impressed.

**MULTILINGUAL AGENTS** Agents who have the ability to speak — and handle calls — in more than one language.

**MULTIMEDIA** In most call centers, a marketing campaign that uses more than one sales or advertising medium. For example, you could combine direct mail with telemarketing for a multimedia sales effort. Or advertise in print, on television and on radio for multimedia advertising. Multimedia is also used for a computer or software that is capable of handling text, images and recorded sound. Last but not least, multimedia is being used by some consultants and market research companies to describe using many different types of technologies (computers, telephones, fax, etc.) within a single business unit. They use it as a catch-all phrase which includes call centers.

**MULTIPLE OUTBOUND CALLING CAMPAIGNS** A single call center can run more than one sales or promotional campaign. Many predictive dialers can separate agents into groups and feed them calls placed from different lists of numbers, and still track their progress for reporting purposes. (Much the same thing can be done for inbound, with ACDs.)

**MULTI-STAGE QUEUING** In ACDs, it is the ability to array a number of agent groups in a routing table. The notion of multiple agent groups being addressed means that the system "looks back" and "looks forward" as it searches for a free agent in the right group to take the call presently holding. Instead of the routing scheme going on to the secondary group and never again considering the primary group, it will keep looking back for an available agent in that first group.

**MUSIC-ON-HOLD** Background music heard when someone is put on hold, letting them know they are still connected. Some modern phone systems generate their own electronic synthesized music. Most phone systems have the ability to connect any sound-producing device, that is a radio or a cassette player. Most companies, unfortunately, devote little attention to the sound source they select. Sometimes competitors will deliberately advertise on the radio station that callers will hear on hold. For many reasons, it is better to use either pre-recorded music that already has clearance from the musicians' rights clearing houses, ASCAP and BMI, or to play a mix of music and messages on hold that are recorded specifically for your company. See MESSAGES-ON-HOLD.

**MUTE** A feature which disconnects the headset microphone so that side conversations (assistance from another agent or supervisor), sneezes or coughs won't be heard by the party on the other side of the line.

**MVIP** Multi-Vendor Integration Protocol. MVIP is a family of standards to let telephony products from different vendors inter-operate within a computer or group of computers. It lets product and application developers put printed circuit cards from different manufacturers (often performing different functions like fax, voice processing and voice recognition) into the same computer or system of computers and have them all work together.

The MVIP Bus was defined by Natural MicroSystems, Natick, MA with assistance from Mitel, Promptus Communications and Rhetorex as a vendor-independent means of connecting telephony devices within a computer chassis. MVIP was introduced by a seven company group in 1990 and has been distributed by Natural MicroSystems, Mitel and NTT International (part of the Japanese telephone company).

MVIP now has several hundred participating companies including companies manufacturing telephone line interfaces, voice boards, FAX boards, video codecs, data multiplexers and LAN/WAN interfaces. MVIP now has its own trade association — the Global Organization for MVIP, which develops extensions such as higher-level application programming interfaces and multi-chassis switching.

**NAME-REMOVAL SERVICE** A service of the Direct Marketing Association (DMA) that lets consumers request to be taken off all mailing lists. It's part of their Mail Preference Service, which also lets people add their names to lists.

**NANP** The North American Numbering Plan. It assigns area codes and sets rules for calls to be routed across North America. (It includes the US, Canada and most of the Caribbean nations). See NORTH AMERICAN NUMBERING PLAN.

**NAP** Network Action Point. An AT&T term describing the switching point through which a call is processed. The NAP switches the call based on routing instructions received from the Network Control Point.

**NAPI** Numbering/Addressing Plan Identifier.

**NARROWBAND ISDN** Quite roughly, any ISDN speed up to 1.544 Mbps.

**NARROWBAND SIGNAL** Any analog signal or analog representation of a digital signal whose essential spectral content is limited to a voice channel of nominal 4 kHz bandwidth.

**NASC** Number Administration and Service Center. The organization that administers toll-free number assignments and the national toll-free number database. Right now the NASC is a division of BellCore, with administrative help from a division of Lockheed.

**NATIONAL CARRIER** Once upon a time (before 1984) the USA had a national telephone carrier known as the Bell System. The law of the land said only AT&T and the Bell System could carry telephone traffic in the United States, with very few exceptions and even fewer exceptions that weren't at the convenience of the Bell System. The USA no longer has a national telephone carrier, but many other countries around the world do. Even when deregulation and liberalization are introduced (as they have been in Europe), as long as a carrier gets some special benefit from national law, you can consider it a national carrier. As the transition to a competitive marketplace occurs, you can call the national carrier the "incumbent carrier."

**NATIONAL CHANGE OF ADDRESS SERVICE** See NCOA.

**NATIONAL DATABASES** Lists of names, addresses, phone numbers and other demographic details for nearly every household in the country. Most national files contain in the neighborhood of 90 to 100 million households. They are compiled from many sources public and private, including customer lists, motor vehicle registrations and warranty cards. They (or segments of them) can be purchased for mailing lists, telemarketing or other marketing plans that require detailed personal or demographic information.

**NATIVE SIGNAL PROCESSING** The concept of using the spare processing power on general purpose microprocessors (found in PCs and often made by Intel) to perform the manipulation of digital signals (echo cancellation, call progress monitoring, voice processing) that is usually done by a specialized microprocessor called a digital signal processor (DSP).

**NC** Network Computer. Similar to a "dumb terminal."

**NCA** Number of calls abandoned. An ACD statistic. The number of calls accepted into the ACD on the trunks only but lost before being connected to a person.

**NCC** Network Control Center. A central location on a network where remote diagnostics and network management are controlled.

**NCH** Number of calls handled. An ACD statistic. A count of all calls handled by a position.

**NCOA (NATIONAL CHANGE OF ADDRESS) SERVICE** A program of the US Postal Service that provides information on changes of address on a national level. Running a list against this data is a big task. Most companies hire a service bureau to do it for them.

**NCR** Was once National Cash Register, which got swallowed up by AT&T who wanted a better computer business, although NCR was best known for its automatic teller machines (ATMs). It was spit out again in 1995 when AT&T broke up (again). NCR is the computer part of what used to be AT&T. See AT&T and LUCENT.

**NET STAFFING** The actual number of agents in a call center minus the required number of agents in a given period. Net staffing that is positive indicates overstaffing; net staffing that is negative, understaffing.

**NET STAFFING MATRIX** A report that shows the actual number of agents, required number of agents, and net staffing for each period of a given day.

**NETWORK** A network ties things together. Computer networks connect all types of computers and computer related things — terminals, printers, etc. A network can cover the entire country, as does the Public Switch Telephone Network, or just a few hundred feet, such as a Local Area Network. Networks are not limited to the call center industry (although they are certainly vital within it). There are train networks and networks of scientists consulting on a cure for cancer.

**NETWORK ACD** Network ACD lets ACD agent groups, at different locations (nodes), service calls over the network independent of where the call first entered the network. NACD uses ISDN D-channel messaging to exchange information between nodes.

**NETWORK INTERQUEUE** Software from Aspect Communications that enables users to network and operate multiple remote call centers as a single site.

**NEURAL NETWORK** A massively parallel computer network that mimics the human brain. It has the ability to learn patterns in and relationships between data. A form of artificial intelligence that is often used in problem resolution software.

**NHLD** Number of calls held. An ACD statistic. The number of calls that waited for a certain amount of time (you decide how long, usually in seconds) before being connected to an agent (or hanging up).

**NIGHT SERVICE** An ACD feature that allows you to specify some alternate performance parameters for after-hours calls. That could be the playing of a particular message telling people when to call back, or it could be a table of alternate routing plans sending calls to another center.

**NNX** A three-digit code that used to identify the local central office. Today the second digit can be any number from 0 to 9.

**NOC** Number of out calls. A call center statistic. The number of outgoing calls made. Also, Networks Operations Center, a group which is responsible for the day-to-day care and feeding of a network. Your long distance carrier probably has one.

**NOISE CANCELLING** Headset manufactures have long sought to reduce the background noise transmitted via headsets. One approach is the use of noise cancelling microphones. These microphones consist of two separate microphones, one directed at the headset user's mouth, the other in the opposite direction. The room side element will pick up ambient room noise along with some ambient user sound. The microphone directed at the user will receive the same amount of ambient room noise as the other microphone, but a much greater amplitude of the user's voice. Both signals are then transmitted to the amplifier. At this point, signals common to both microphones are cancelled out. What remains is the extra voice signals received by the user side microphone. This signal is then amplified and transmitted to the party on the receiving end of the call. This approach has one drawback. It demands perfect microphone positioning, because without it, the headset user's voice is cancelled. The technology works well with highly-trained people such as pilots, astronauts, and military personnel, but can be difficult to implement in the office environment where less skilled personnel struggle to properly position sensitive microphones. Headset manufacturers compromised by using noise cancelling microphones with more limited capabilities but that were easier to use.

A second approach to noise reduction is the use of voice switching technology. This technique only allows the microphone to transmit when volume reaches a predetermined level. When the headset user is not talking, or is pausing during the conversation, no sound is transmitted. When the headset user speaks at a normal level, the microphone is "live" and will transmit in a normal fashion. This approach also has it drawbacks. When the microphone is "live" it picks up not only the voice of the person using the headset, but any and all background noise. Voice switching helps the headset user hear what is being said more clearly, but does little to help the person to whom they are talking.

As a solution, some headset manufacturers have merged the two technologies. By using a noise cancelling microphone and voice switching, they achieve near perfect noise reduction. Each manufacturer offers noise cancelling technology on some of their headsets.

Noise cancelling is important in a telephone call center. In a large center, as room noise rises, agents speak louder. For those employees, noise is more than just an inconvenience, or a black spot on a professional image, it directly affects productivity. When conversations must be repeated, call durations increase. Multiply this by enough calls, and staffing and equipment must also be increased. The above information from headset distributor, CommuniTech.

**NON-APPARENT SOURCES** Something based on an assumption, rather than a direct factual link. For example, when you are trying to find phone numbers that correspond to names on a mailing list, you might find a match based on a past address or an ambiguous abbreviation. Sources that can't be directly tied to the person, like a Social Security number, customer account number, or direct address, are said to be non-apparent.

**NON-PUBLISHED** A telephone line with no phone number listed in the telephone company's directory. It is different from an "unlisted" number, which is a consumer phone number not listed in the telephone directory at the request of its owner. As we understand it, "non- published" is the broader term covering unlisted numbers, plus business telephone numbers such as Centrex numbers, or secondary trunks, which are simply not listed anywhere, even in the telephone company's directory.

**NORMALIZE** To change an unusual call statistic reported by the ACD to reflect what would have been usual for that period of the day or that day of the week. You would do this before entering the statistic into your collection of historical data, so the historical patterns will not be distorted by the unusual data.

**NORTH AMERICAN NUMBERING PLAN** NANP. The format for assigning and administering telephone numbers in North America. It consists of the three digit area code, a three digit exchange or central office code and a four digit subscriber code. The latest version of this plan, introduced in January 1995, allows the second digit of the area code to be any digit from 0 to 9. This will allow more than 6 billion telephone numbers and 792 area code combinations.

**NORSTAR** A family of telephone systems from Nortel. Significant insofar as it was one of the first small business phone systems open enough for developers to create add-on applications - like ACDs - to run on top of it, enabling the development of the small or departmental call center. Still a lot out there, but not much of a factor in the industry anymore.

**NORTH AMERICAN AREA CODES** See NPA and NANP.

**NORTH AMERICAN NUMBERING PLAN** See NANP.

**NPA** Numbering Plan Area. A fancy way of saying "area code." There are 792 available area codes to serve the United States, Canada, Bermuda, the Caribbean, and Northwestern Mexico. The NPA is the first three digits of a ten digit telephone number. It is required for dialing long distance calls. In an area code, no two telephone lines may have the same seven digit phone number.

These days the first digit of an area code is a number from 2 to 9 (known as "N" in telecom talk) and the second and third digits can be any number ("X"). Central office, or exchange, codes use the same pattern. Switching systems in the national network differentiate between the central office and area codes by recognizing the subscriber always dials 1+ or 0+ preceding an area code when direct dialing long distance calls.

Some area codes are not used to designate a geographical region; they are used for special purposed. Here are the special, unassigned and reserved NPAs:

200 Reserved for special services

211 Assigned to local operators

300 Assigned to special services

311 Reserved for special local services

400 Reserved for special services

500 Reserved for special services

511 Reserved for special local services

600 Reserved for special services

700 Assigned special access code for interLATA carriers/resellers

711 Reserved for special local services

800, 822, 833, 844, 855, 866, 877 and 888 are reserved for current and future toll-free (or In-WATS) use

900 Pay-per-call services

911 Local emergency services

**NTH NAME SELECTION** Let's say you wanted to sample a list. How would you pull the names? One of the simplest methods would be to select every tenth or 34th name. This technique is called "Nth name selection." "N" is the number you choose.

**NUMBER ADMINISTRATION AND SERVICE CENTER** See NASC.

**NUISANCE CALLS** Automated dialers often place more phone calls than there are agents to handle them. This is because the dialer is trying to screen out the unproductive calls and make sure the agent is always talking to someone, not waiting. But the dialers don't always guess right, and sometimes calls are placed that get through, and nobody is one the other end. Those are the "Hello? Hello?" calls everybody hates — nuisance calls. With predictive dialers, the nuisance call rate can be reduced to as low as 1%. Call center managers have to set a balance between how productive they want their agents to be and how high is the acceptable nuisance level.

**NUMBER OF CALL ABANDONED** See NCA.

**NUMBER OF CALLS HANDLED** See NCH.

**NUMBER OF CALLS HELD** See NHLD.

**NUMBER OF OUT CALLS** See NOC.

**NUMBER-FINDER** A company that takes lists of names and/or addresses and roots out phone numbers for them. Techniques include using computers to search huge databases, making calls to directory assistance, and looking numbers up in phone books.

**NUMERIC KEY PAD** A separate section of a computer keyboard which contains all the numerals 0 through 9. Sometimes, some special keys are included — a plus sign, a minus sign, a multiplication sign and a division sign. The numeric key pad on a computer is the same as that found on calculators and adding machines. The top row is 789. The second top row is 456. The third top row is 123. The lowest row is typically 0, "." and "+". The numeric key pad is exactly opposite that of the touchtone telephone keypad, which was designed deliberately to be unfamiliar to users, so they may not input digits into the nation's telephone system faster than it could take them. Early touchtone central offices were very slow.

**NXX** N represents any digit from 2 to 9 and X is any digit. This represents the numbering scheme for both the area code (first three digits in a ten-digit, long distance, telephone number) and the exchange or central office code (the second three digits in the same number). The telephone network tells the difference between the two by the "1" or "0" dialed before the number that shows it is a long distance call. (Which is why neither can have a 0 or a 1 as its first digit.)

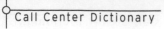

**OAI** Open Application Interface. An opening in a telephone system that lets you link a computer to that phone system. It lets the computer command the phone system to answer, delay, switch, hold etc. calls.

Typical applications include simultaneous data file and call delivery (such as in a call center where a client record is pulled up with ANI and delivered along with the call), message desks where a single telephone switch serves more than one company and predictive dialing. Also known as switch-to-host interface and, when the telephone switch is a PBX, PHI, PBX-Host Interface.

In the early 1990s, OAI was an important concept through the telecom industry, but it is particularly important to call centers. Among the first OAI applications were call center applications. The economy of scale in large call centers, where shaving a few seconds of each call results in thousands or millions of dollars in savings, provided the first market of OAI applications. That economy of scale continues to make OAI applications attractive to call centers of all kinds.

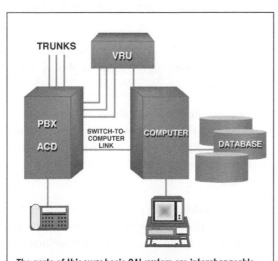

The parts of this very basic OAI system are interchangeable. The PBX/ACD could be a predictive dialer, a PBX, an ACD or even a key system. The computer can be a mainframe, a mini-computer or any type of LAN. The database could include account information, order processing info, sales info or just about any other data.

Before OAI the link between the telephone system and the computer system was the agent, sitting at a desk with a workstation and a station set. And it certainly seems that every call center application requires access to a database. The old way was pretty hard on the agent, and wasted time. With OAI the computer system and the telephone system are tied together for the benefit of the agent. Agents work more effectively and serve customers better.

Another reason OAI is important to call centers is because it allows you to create your own telephone system applications, hire a developer to create the application of your dreams or buy a package created for your niche by a developer. Your call center is unique. Call center managers are constantly tweaking their technology. OAI allowed the wealth of apps we now have at our disposal to be created in the first place.

**OBJECT-ORIENTED PROGRAMMING** A method of creating computer software that uses building blocks of data and instructions, called "objects" instead of single lines of code. The idea of object oriented programming is make the writing of complex computer software much easier, to simply combine objects together to produce a fully-written software application. If a vendor touts "object-oriented programming" in a call center software, it is trying to tell you that it will be easy to change. In development software, it is trying to tell you that it is easy to use.

**OBJECTIONS** When a salesperson asks a prospect for an order the prospect may say "no." All good (and even most not so good) salespeople then ask "Why not?" The reasons the prospect gives (real or imaginary) for not buying are his or her objections.

**OCC** An abbreviation for occupancy. Used in ACD reports to show the percentage of time agents are at their stations and ready to receive calls. Also see OTHER COMMON CARRIER.

**OCCUPANCY** The percentage of the scheduled work time that employees are actually handling calls or after-call wrap-up work, as opposed to waiting for calls. The amount of time your agents are sitting in their seats, handling calls and doing follow-up work. The percentage itself is often known as the "occupancy rate." The term is also used for switches and circuits in addition to agents. "Occupancy" is the time a circuit or a switch is in use.

**OCDD** On-line Call Detail Data. An AT&T method of accessing ANI (Automatic Number Identification) information from their computer within 48 hours after receiving the telephone call.

**OCTOTHORPE** An extremely obscure term for the character at the bottom right of your touchtone keypad, which is also known as the "pound sign" or the "number sign." Some phone systems use it to represent "no," others to represent "yes." It is commonly used for special functions in telephone systems, voice processing systems and long distance telephone networks.

**ODBC** ODBC Open Database Connectivity is Microsoft's strategic interface for accessing data in a heterogeneous environment of relational and non-relational database management systems. Based on the Call Level Interface (CLI) specification from the SQL Access Group, ODBC provides a vendor-neutral way of accessing data in a variety of personal computer, minicomputer and mainframe databases.

**OFF-HOOK** When the handset is lifted from its cradle it's "off-hook." Lifting the hookswitch on a single line telephone alerts the central office that the user wants the phone to do something like dial a call. A dial tone is a sign saying "Give me an order." The term "off-hook" originated when the early handsets were actually suspended from a metal hook on the phone. Some phones have autodialers in them. Lifting the phone signals the phone to dial that one number. An example is a phone without a dial at an airport, which automatically dials the local taxi company. All this by simply lifting the handset at one end — going "off-hook."

**OFF-NETWORK** In this term "network" usually refers to a private, corporate network or a long distance network. If you use a small long distance company you are charged a different rate for calls that are off-network. A call to Illinois would be off-network if your long distance company serves mostly or only Michigan. (These types of long distance companies do exist.) If your corporate network links major company locations, anything outside those locations would be off-network. "Off-net" calls are more expensive than network calls.

**OFF-PEAK** Times outside a call center's busy, or peak, periods. Often centers will have different procedures for handling off-peak calls. Breaks may be scheduled only for off-peak times, call flows may be changed completely, or calls may be routed to a voice processing system.

"Off-peak" is also the after-business-hours period in which telephone carriers offer lower rates. If you do outbound telemarketing to consumers, you get to take advantage of these rates by calling your prospects at home during off-peak hours.

**OFF-PREMISE EXTENSION** OPX. A telephone located outside of your office building, but attached to your office telephone switch via a dedicated telephone line. The off-premise phone can use all the features of the telephone switch. Off-premise extensions are commonly used by answering services.

**OFF-THE-SHELF** Usually used for software, but sometimes also used to describe other call center systems. It is a system that is not customized, but sold to all comers as-is. You do any possible customization yourself.

**OFFER SCRIPT** A term for campaign-specific text that appears on the agent's screen, guiding them through the transaction.

**OFFERED CALL** A call that is received by the ACD. Offered calls are then either answered by an employee (handled) or abandoned. Another meaning for "offered call" is a call that is presented to a trunk or group of trunks. See TRAFFIC ENGINEERING.

**OFFERED-TO-SWITCH** The ACD that gets the call sent by the "offering switch" for possible handling.

**OFFERING SWITCH** A term that refers to an ACD that offers an arriving call to another ACD for possible handling. The offering switch does NOT give up control of the arriving call unless the offered-to switch indicates that it can handle the call.

**ON-HOOK DIALING** Allows a caller to dial a call without lifting the handset. The caller can listen to the progress of the call over the phone's built-in speaker. When you hear the called person answers, you can pick up the handset and speak or continue to use the speakerphone.

**ON-LINE** A descriptive word used for computer operations, such as information access and processing. An on-line inquiry system offers real-time access to a database of information. On-line processing takes place on the fly. As opposed to batch processing, where a bunch of functions are performed at once.

**ONE WAY TRADE** A schedule trade in which only one employee is working the other's schedule.

**OPEN APPLICATION INTERFACE** See OAI.

**OPEN ARCHITECTURE** The ability of a hardware platform to be built out, usually by third-party developers, to add features and new uses. This is most often used to describe the openness (or closed-ness) of PBXs and ACDs. This term was a replacement for the more PBX-specific term OAI as the telecom industry progressed through the 1990s. In general, though, any hardware platform, including computers, and many software platforms (especially operating systems) can be described as being of an open or closed architecture.

**OPEN COLLABORATION ENVIRONMENT** OCE. Apple's Open Collaboration Environment extends the Macintosh operating system to provide a platform for the integration of fax, voicemail, electronic mail, directories, telephony and agents.

Let us put this in as clear a way as we can without being harsh: Apple and its platforms are not competitive or used widely in the call center industry.

**OPEN NETWORK ARCHITECTURE** ONA. The FCC's idea to promote value-added telephone services (voice mail, electronic mail, shopping by phone) without creating a monopolistic mess is called ONA — Open Network Architecture. Under this concept, the telephone companies are obliged to provide a certain class of service to their own internal value-added divisions and the SAME class of service to nonaffiliated (that is outside) valued-added companies. The concept is that the phone company's architecture is to be "open" and that everyone and anyone can gain access to it on equal footing.

**ONA** requirements were imposed on GTE in early 1994. Subsequent FCC orders have substantially reduced the applicability of unbundling and other aspects of ONA on AT&T. Currently, AT&T is not directly subject to ONA requirements, but is subject to Comparatively Efficient Interconnection (CEI) requirements. (Visit www.fcc.gov if you want to know what that is.)

**OPERATING TIME** The time required for seizing the line, dialing the call and waiting for the connection to be established.

**OPERATOR** A term used by the general public interchangeably with the term "agent" or "receptionist." In the telecommunications industry, "operator" is generally reserved for employees of a telephone carrier. They offer directory assistance and other special services. The person who answers the telephone for a business is an "attendant." A person who works in a call center is an agent, a representative or goes by some specialized name.

**OPERATOR SERVICES** OS. Any of a variety of telephone services which are offered through the assistance of an operator, either human or automated. Automated operators include using interactive voice response and speech recognition. These services include collect calls, third party billed calls and person-to-person calls. There are companies that provide these operator services to smaller long distance carriers for a fee. One operator service bureau may even provide service for several carriers. These operator service bureaus are an excellent example of an inbound call center.

**OPERATOR WORKSTATION** OWS. The OWS is an advanced voice and data workstation (typically a PC running a flavor of Windows) that streamlines and automates many of the routine tasks of an operator, thus reducing the amount of time needed for call handling. Color screens, pop-up windows, one-touch commands, and database look-up are some of the features that simplify the operator's tasks and speed call processing.

**OTHER COMMON CARRIER** OCC. This is any long distance carrier OTHER than AT&T. AT&T's tariffs have been loosened considerably in the last few years, so this term is becoming less important. Once it was important to note the difference between AT&T, with its highly restricting tariffs, and its competitors that had much fewer restrictions. Now the difference is less great and the term is used less. To refer to all long distance carriers including AT&T "IXC (Inter-eXchange Carrier)" is used.

**OUT OF BAND SIGNALLING** A method of controlling information in a telecommunications network using signals that are carried in a band or channel that is separate from the band carrying the information. The information would be, for example, the voices in a telephone call. Out of band signalling is much less prone to tampering than in-band signalling, and allows for many features in-band signalling can't provide, such as caller identification (receiving your caller's phone number before you pick up the phone). See SIGNALING SYSTEM 7.

**OUTBOUND** In this dictionary, a term used to describe calls. Outbound calls are made by your company to other people, off your site. From your point of view (or the point of view of your company), the are calls leaving, or going out, hence "outbound."

**OUTDIALING** In telemarketing, initiating calls is called outdialing. It can be done with or without operator supervision, and with a varying degree of technological supervision. Outdialing is an umbrella term that refers to any of the common dialing methods: preview dialing, predictive dialing, or power dialing.

**OUTDIAL FACILITY** A term for a group of trunks available for placing outbound calls by dialing a user-defined access level. Synonyms include trunk queuing group.

**OUTGOING TRUNK** A line or trunk used to make calls.

**OUTGOING WATS** An outgoing WATS (OutWATS) trunk can only be used for outgoing bulk-rate calls from a customer's phone system to a defined geographical area via the dial-up telephone network. Originally WATS lines came in only lines that could receive calls or lines that could make calls. Now, you can buy a WATS line that handles both incoming and outgoing lines. See WATS.

**OUTSOURCING** Contracting one (or more) of your company's internal functions (help desk, telemarketing, payroll) to an outside company. A familiar example is ADP, a company that handles payroll for many businesses. When your company hires and inbound or outbound telemarketing service bureau, it is also outsourcing.

Companies outsource because they do not want to risk capital on a new enterprise (such as testing telemarketing), because they don't have the expertise or physical resources to do the job right (often the case with help desks), because they believe a third party can do the job more cheaply, or because they want to concentrate their resources on what they feel they do best.

The drawbacks to outsourcing are similar to the drawbacks to renting as compared to buying. With outsourcing your company never gains expertise in the function. You are always reliant on the expertise of your vendor. You don't have the tax advantages of depreciation, nor does the high tech equipment your fee is paying for ever appear under assets on your balance sheet. None of these drawbacks are standing in the way of a general trend toward outsourcing.

**OVERFLOW** Additional traffic beyond the capacity of a specific trunking group, agent group, telephone system or call center. This traffic can be offered to another trunk group, agent group, switch or center.

**OVERFLOW CALLERS** Calls that come in to an ACD that are beyond the capacity of the available agents. These are three options for these calls: 1) Put them on hold until someone becomes available, 2) Let them leave a message and call them back, and 3) Route them to an alternative group of agents.

**OVERFLOW CAPABILITY** When calls come into a call center via an ACD, they are routed to an agent (or an agent group) based on the internal call distribution tables you've preprogrammed. When you get more calls than you have agents, that's "overflow." An ACD can keep those callers on hold until someone becomes available. But if it has overflow capability it can send them to another group of agents (at another location, or attached to a different campaign).

**OVERFLOW GROUP** A secondary group of agents assigned to receive a certain type of call when all the agents in the primary group are busy.

**OVERFLOW LOAD** The part of an offered load that is not carried. Overflow load equals offered load minus carried load.

**OVERFLOW TIE-LINE ENHANCEMENT** Using Overflow Tie-Line Enhancement, non-ISDN calls diverted to an overflow call center now convey the city-of-origin announcement prior to being connected to an agent. This announcement should tell the agent what type of call is being handled, especially if the original call center handles slightly different calls than the overflow call center.

**OVERFLOW TRAFFIC** The part of the offered traffic that is not carried, for example, overflow traffic equals offered traffic minus carried traffic.

**OVERLAY** Adding information from one list to another list. For example, enhancing your house list with demographic info from a general consumer list. Overlay also means having one area code designation placed atop another one, as often happens in cities when landlines get one area code but cellular phones and pagers get a different one.

**OVERLOAD CONTROL** How a system responds to being overstressed is called "overload control." When a system is overloaded, frequently there are so many extra events being processed that the system's actual capacity or throughput goes down. Even though it may be rated at, say 10,000 busy hour calls, when overloaded, for example, with 11,000 calls, the computer telephony system may be only able to process only 8,000 calls.

**OVERLOAD MANAGEMENT** A term for handling peak call demands by selectively delaying, degrading or dropping only those portions of traffic flow that are tolerant of those particular types of impairments.

**OVERLOAD PROTECTION** An uninterruptible power supply (UPS) or power protection term. Refers to automatically shutting the unit off when overloaded to protect against overload damage.

**OWS** See operator workstation

**PABX** Private Automatic Branch eXchange. Originally, PBX was the word for a switch inside a private business (as against one serving the public). PBX means a Private Branch Exchange. Such a "PBX" was typically a manual device, requiring operator assistance to complete a call. Then the PBX went "modern" (i.e. automatic) and no operator was needed any longer to complete outgoing calls. You could dial "9." Thus it became a "PABX." Eventually, all PBXs acquired these automatic features and the "A" became irrelevant. So the PABX is commonly referred to as a PBX, and PABX is an obsolete term. (See PBX for more information about what they do and where they fit into the call center.)

**PACING ALGORITHM** Predictive dialers use complicated software techniques to achieve a careful balance between the number of available agents and the number of calls placed. The pacing algorithm is the mathematical formula the dialer uses to decide how many calls to place at any given moment, taking into account such factors as the average length of calls, the time of day, the accumulated information on the agents' speed, and the hit rate — how many calls it can complete versus how many busy and unanswered calls it places. Essentially, the predictive dialer is its pacing algorithm. That is the key component that differentiates one system from another.

**PAID HOURS** The time that an employee is either on duty- handling calls, doing other work, in meetings, etc., or on a paid schedule exception, such as an excused absence. In TCS's TeleCenter System, for example, this is calculated as scheduled hours minus any unpaid schedule exceptions that occur within those scheduled hours.

**PAPM** Primary average positions manned. An ACD statistic. The average number of positions manned within a defined period whose primary job is to answer calls directed to that group.

**PASSIVE BUS ISDN** feature which allows up to six terminal devices and two voice devices (also called telephones) to simultaneously share the same twisted pair, each being uniquely identifiable to the switched ISDN telephone network. See ISDN.

**PASSIVE PROGRAM** An entertainment or information program delivered by phone (usually as part of a pay-per-call service) that doesn't incorporate menus, voice prompts, or any other interactive choices. It is a plain vanilla call. It is used for promotions that target a fair number of rotary phones. Very similar, if not interchangeable, with audiotex.

**PAY-PER-CALL** A service that charges the caller for information provided above the simple cost of the phone call. It refers to 900 numbers (and the local 976 or 540 versions). Pay- per-call has limited application in the call center, though not for lack of trying. There have been many attempts in the last decade to create momentum for business uses of 900. People have tried putting professional services (like lawyers or tax advice) on a pay-per-call basis, but the response was underwhelming. The problem with 900 is that businesses tend to block access to that code in their phone systems, putting any service at a disadvantage. Even pay-per-call customer support, which for a time looked like the most promising application, never really took off. (If you want to have someone pay for support, it's easier to take a credit card and give them an 800 number to call.)

A few years ago, AT&T came out with an interesting twist on pay- per-call called Vari-A-Bill, which allows a service provider with the proper equipment to change the cost of a call in mid-stream, letting the provider set up an IVR-like menu of possible options for a caller. It's great for service bureaus that offer pay-per-call as one of many service options.

**PBX** Private Branch eXchange. A variety of business phone system. It's a smaller version of the phone company's larger central office switch. The difference, besides size, is that you own the PBX, and it sits in your office, not theirs.

The thing that gives a PBX its basic nature is this: users have the ability to dial out by themselves, without the intervention of an operator to patch the call through for them.

At the time of the Carterfone decision in the summer of 1968, PBXs were electro-mechanical step-by-step monsters. They were 100% the monopoly of the local phone company. AT&T was the major manufacturer with over 90% of all the PBXs in the U.S. GTE was next. But the Carterfone decision allowed anyone to make and sell a PBX. And the resulting inflow of manufacturers and outflow of innovation caused PBXs to go through five, six or seven generations — depending on which guru you listen to. By the fall of 1991 PBXs were thoroughly digital, very reliable, and very full featured. There wasn't much you couldn't do with them. They had oodles of features. You could combine them and make your company a mini-network. And you could buy electronic phones that made getting to all the features that much easier. Sadly, by the late 1980s the manufacturers seemed to have finished innovating and were into price cutting. As a result, the secondary market in telephone systems was booming. Fortunately, that isn't the end of the story. For some of the manufacturers in the late 1980s figured that if they opened their PBXs' architecture to outside computers, their customers could realize some significant benefits. (You must remember that up until this time, PBXs were one of the last remaining special purpose computers that had totally closed architecture. No one else could program them other than their makers.) Some of the benefits customers can realize from open architecture include:

• Simultaneous voice call and data screen transfer.

• Automated dial-outs from computer databases of phone numbers and automatic transfers to idle operators.

142

- Transfers to experts based on responses to questions, not on phone numbers.

An alternative to getting a PBX is to subscribe to your local telephone company's Centrex service. For an explanation of Centrex and its benefits, see CENTREX. Here are some of the benefits a PBX has over Centrex:

1. First and foremost: you can build a call center on a PBX, but building one on Centrex will give you nothing but headaches — if you can do it at all.

2. Ownership. Once you've paid for it, you own it. There are obvious financial and tax benefits.

3. Flexibility. A PBX is a far more flexible than a central office based Centrex. A PBX has more features. You can change them faster. You can expand faster. Drop another card in, plug some phones in, do your programming and bingo you're live.

4. You can still have Centrex, too. You can always put Centrex lines behind a PBX and get the advantages of both. In some towns, Centrex lines are cheaper than PBX lines. So buy Centrex lines and put them behind your PBX. Make sure you don't pay for Centrex features your PBX already has.

5. PBX phones. There are really no Centrex phones — other than a few Centrex consoles. If you want to take advantage of Centrex features, you have to punch in cumbersome, difficult-to-remember codes on typically single line phones. PBXs have electronic phones, with screens and dedicated buttons. They're usually a lot easier to work. It's a lot easier to transfer a call, for example, or create a conference between users, etc. The whole experience is a lot more productive for the average user.

6. Voice Processing/Automated Attendants. Centrex's DID (Direct Inward Dialing) feature was always pushed as a big "plus." You saved operators. However, you can now do operator-saving things with PC-based voice processing and automated attendants you couldn't do five years ago. These things work better with on-site standalone PBXs than with distant, central office based Centrex. Moreover, virtually every PBX in existence today supports DID. You can dial directly into PBXs and reach someone at their desk just as easily as you can dial directly using Centrex.

7. Open Architecture. Most PBXs have open architecture. Central offices don't.

8. Good Reliability. There have been sufficient central office crashes and sufficient improvement in the reliability of PBXs that you could happily argue that the two are on a par with each other today. Both are equally reliable, or unreliable. The only caveat, of course, is that you back your PBX up with sufficient batteries that it will last a decent power outage. Of course, that assumes that your people will be prepared to hang around and answer the phones during a blackout.

Which brings us to the call center. When most people think of call centers, they think of massive rooms filled with people, tied together by a mammoth ACD network and thousands of lines. That's the popular picture, but it's not the whole picture. Small call centers have to start somewhere. Small businesses outnumber large ones. Small call centers outnumber large ones as well. The fastest, easiest way to start a call center at your business is to take the people who answer calls, take orders, call customers, etc., and link them on a data network. Then, add ACD features to your PBX using one of many recently developed software applications, and you can simulate the big-business look and feel with small-business flexibility. Building call centers out of souped up PBXs is one of the fastest growing segments of the call center market.

Those PBX/ACDs may not meet your every call center need, but they can tide you over, or let you create a departmental center that you can use as a test bed.

**PBX/ACD** A PBX with automatic call distributor (ACD) features. This arrangement can work well for smaller call centers. It's also a great way to try out the idea of an ACD or call center. If your PBX has this feature, use it. You won't have to pay any more to get it. Just set it up.

An ACD takes a lot of processing power. If your call center grows too large (just how large will depend on your PBX and a host of other things), the ACD feature can bog things down for the whole system. Compare to ACD.

**PBX EXTENSION** A telephone line connected to a PBX.

**PBX FRAUD** Same as TOLL FRAUD.

**PBX INTEGRATION** A loose term that means joining the PBX to any number of outside computer-based gadgets and services, from voice mail to call accounting. For example, if you want to integrate voice mail into a PBX, you minimally need this: the ability to provide a message waiting indicator (light or stutter dial-tone) at the user's phone when a message is received, and to forward a call to the user's mailbox when a call is sent to the recipient and they are on the phone or don't answer (forward-on- busy or ring-no-answer). This requires PBX "integration."

PBX integration data may be implemented in-band or out-of-band on a separate link, most often a serial link. Some PBXs "integrate" with outside equipment better than others.

**PBX PROFILES** PBXs do things differently. To make a conference call, one PBX's phone may put the caller on hold automatically, while another may insist that you put that person on hold manually and then dial the next person to join the conference call. As Novell in the fall of 1994 attempted to get as many PBXs as possible to conform to TSAPI, it discovered that PBX features often work very differently. So it decided to categorize PBXs and their features. They called these PBX profiles.

The idea was that Profile A would contain the most common, easy-to-integrate-to-TSAPI features. B would contain the second most common, etc. Novell also calls PBX Profiles PBX Driver Profiles.

**PC-CENTRIC** There are two ways you can organize a computer to control telephone calls on an office telephone system. One way is to join a file server on a local area network to a phone system. Commands to move calls around are passed from the desktop PC over the LAN to the server and then to the phone system via the cable connection between the server and the system. A second way to get a computer to control phone calls is through a connection at the desktop. This is called the PC-Centric method.

There are two ways you can do this. The first is to join the desktop phone to the computer with a cable. This is often done through the PC's serial port, connected by cable to the phone's data communications port (if it has one). The second way to be PC- Centric is by simply replacing the standalone phone with a board that emulates a phone and dropping it into the PC's bus.

**PCM** See PULSE CODE MODULATION.

**PEAK HOUR** When used with an automatic call distributor, the peak hour is when the number of calls coming into your center are at their highest level. ACDs allow you to track and report on calls by hour. Some allow you to also track peak half-hours, or peak days of the week or months of the year.

**PEAK LOAD** A higher than average quantity of traffic. Peak load is usually expressed as a one-hour period, often the busiest hour of the busiest day of the year. See BUSY HOUR.

**PEAK PERIODS** Times when the number of calls coming into a call center is at its highest level. You can adjust ACDs to respond to peaks by adding more agents to busy agent groups if you use the MIS reports to track them. Or ACD forecasting software to predict when those peaks might occur.

**PEAK POSITION REQUIREMENTS** The maximum number of base staff required in any half hour within a given date range.

**PEER-TO-PEER NETWORK**
1. A data network (typically a local area network) in which every node has equal access to the network and can send and receive data at any time without having to wait for permission from a control node. While peer-to-peer resource sharing is effective in small networks, security and reliability issues prevent its widespread use in larger networks.

2. A telephony term describing the relationship between a telephone system and the external computer working with it. Picture a telephone switch acting as an

automatic call distributor and an outboard computer processor. The idea is to coordinate the call and the screen at the agent's desk. Communication must take place between the switch and the computer. If that communication is peer-to-peer, as it is, for example, in the DMS Meridian ACD, then neither the switch nor the computer is in a "slave" relationship to the other.

**PEG COUNT** A raw count of some event. In the call center context, it's a count of the number of calls placed or received at a certain point or over certain lines during a period such as an hour, day or week. A peg count simply tells you how many calls you made or received. It does not tell you how long they were or where they went or anything else. In the old days before we had accurate and relatively inexpensive call center management systems, we relied on peg counts to figure out how many circuits and agents we needed. Because this data was originally maintained by moving pegs on a board with units of 1s, 10s, 100s, 1000s it became a peg count. It's rarely, if ever, still used for this purpose.

**PENETRATION**
1. The number of names actually on a list compared to the total number of names possible for that list. The trick here is having a good idea what the actual number possible is: some people don't have phones or driver's licenses, for example.

2. In an outbound campaign, the degree to which you have reached (or attempted to reach) all the names on a given list. That is, if you have a list of 100,000 names, and your dialer calls 80,000 of them, you have 80% penetration.

**PERCENTAGE ATB** Percentage of All Trunks Busy. Percentage of time during a reporting period that all trunks in a group or split were busy. This may be measured in two ways, actual simultaneous busies and call length per event, or you can back into it statistically. Neither technique is absolutely accurate as each depends on "snap shots" in a environment of random interleaved call events.

**PERCENTAGE CA** Percentage of Calls Abandoned. Indicates the percentage of calls abandoned by callers after being accepted by the ACD.

**PERCENTAGE HLD** Percentage of total calls HeLD in queue within a reporting group.

**PERCENTAGE NCO** Percentage of total of Number of Calls Offered to a particular reporting group.

**PERCENTAGE TUT** Percentage of Trunk Utilization Time. The percentage of a time during a reporting period that a trunk is in use and not idle.

**PERIPHERAL EQUIPMENT** Equipment not integral to but working with a phone system. An example might be a printer or television screen on which calling traffic statistics are displayed. It might also be a voice mail system. AT&T once called PBX peripheral equipment "applications processors," because they process specific applications. Some people now call them Adjunct Processors or Outboard Processors. Other examples of peripherals include voice response units, announcement systems, message-on- hold players, fax-on-demand servers, and some outdialers.

**PERMANENT SHIFT TYPE** When using workforce management software to create staff schedules, this is a shift definition that the program gives priority to. It uses this type of shift definition only as long as no overstaffing results in any intra- day period. (At which point flexible shift types are used.) When scheduling is done for more than a week at a time, the permanent schedules are always identical from one week to the next.

**PERSONAL 800 NUMBER** Several long distance companies are now offering Personal 800 numbers, which are basically party line 800 numbers with call routing. The way they work is as follows: You dial a number, e.g. 800-484-1000. A machine answers with a double beep. You punch in four or five digits on your touchtone pad. A voice response unit at the other end hears the digits, says "Thank you for using MCI" and dials out your long distance number. The per minute charges are more expensive than normal 800 lines. One company, MCI, is also offering FOLLOW ME 800 which allows you to change the routing of your personal 800 number instantly with one phone line.

This service (and other new toll free services like it) were partly responsible for the chain of events that led to carriers hoarding 800 numbers, and ultimately for the need to create a brand new toll free exchange in 1996 (888) and at least one more following (877).

**PERSONAL INFORMATION MANAGER PIM.** Watered-down contact management software that is less oriented toward sales and more toward automating routine daily tasks. Typically a PIM has a popup calendar, address and phone call management, appointment scheduler and other utilities designed to clear things off your desktop. They are increasingly aware of Caller ID, networks, and electronic messaging.

**PERSONAL IVR** An Interactive Voice Response system running on your own personal PC and designed to serve the needs of only one person. See IVR.

**PHONE APPENDING** The process of attaching a phone number to a name or address record on a list or in a database. This can be done very quickly (and pretty cheaply) by a list and lookup service bureau.

**PHONE PHREAK** The original hackers. Long before the invention of the personal computer, errant youth took joy in messing about with the technology of large corporations. The telephone system, in those days the Bell System, provided a technological challenge to petty criminals, social reformers and bright college students who wanted to make a free call. Phone phreaks developed many devices that helped them fool the System into giving them something for free. (In one case they had help from the makers of Cap'n Crunch cereal, who developed a toy whistle that happened to do exactly what they needed.) These days people who mess with any telephone system (there are many now) are lumped in with the computer hackers/crackers. The tradition lives on.

**PHONEME-BASED** No, it's not "phone-me." It's pronounced "fo-neem." A phoneme is the tiniest unit of speech. It is similar to a single letter in written language, but in English the letter "g" has two sounds — and is represented by two phonemes. There are about 40 phonemes in the English language.

In a phoneme-based speech recognition system, new words are easier to add, because all the phonemes are already programmed in. Adding a new word is a matter of collecting the proper phonemes for the word. The technique is on the developing edge of speech recognition.

**PILOT PROGRAM** If you plan to embark on some ambitious new project (launching a new product, say, or testing out a new marketing strategy), using a pilot program can save you headaches. Especially if your program involves your call center. For example, say you run a bank, and you want to install a new IVR front end, greeting all your customers and offering them options to deal with their account questions. Instead of just throwing money at an IVR vendor and saying "make it so," hoping that it will go smoothly and your customers will love it, start small. A pilot program is where you go to a service bureau and contract for a small test. You use their IVR system, see how you like it, and see if your customers react well. If it works, roll it out nationally. If it doesn't, well, you saved yourself a lot of heartache.

**PILOT SURVEY** A small market research study conducted in advance of the main survey. It is used on a control group of respondents to fine tune the wording of questions and to knock out any that are superfluous.

**PJ-327** A two-prong telephone handset or headset connector for an operator console or ACD agent station.

**PLANT TEST NUMBERS** Virtually every 800 IN-WATS number has a plant test number. This is its equivalent seven digit local number, with a standard three-digit central office exchange code and a four-digit extension.

The purpose of plant test numbers is to allow the telephone company to test the local part of the incoming 800 number by simply dialing that number. For example, Miller Freeman has an 800 number — 800-LIBRARY (or 800-542-7279). The plant test number of the first line of that 800-LIBRARY group is 212-206-6870. It is valuable to know the plant test numbers of your incoming WATS lines so you can test the local loop part of those lines. The local loop part is the part which typically gives the most problems. It is, unfortunately, the only part of your 800 lines you can test yourself — unless you ask someone (or several people) to call you regularly on your 800 lines, just to test them. You can get plant test numbers out of your local and/or your long distance carrier. When they tell you those numbers are "not available," beg a little. They are available and you are entitled to them. Calling plant test numbers costs exactly what a normal long distance IN-WATS call on that line costs. So keep your test calls short. You should call your plant test numbers once a day.

**PLENUM** The space between a drop ceiling and the real ceiling. In office buildings this space is used to circulate air. If you want to run cable in this space it must meet strict fire resistance standards.

**POINT OF PRESENCE (POP)** A POP is the place your long distance carrier, called an IntereXchange Carrier (IXC), terminates your long distance lines just before those lines are connected to your local phone company's lines or to your own direct hookup. Each IXC can have multiple POPs within one LATA. All long distance phone connections go through the POPs.

**POINT OF PURCHASE (POP)** The retail store or other location where the final customer or end-user buys your product. Sometimes used to refer to the retail shelf or display where the customer selects the product.

**POINT OF SALE (POS)** Similar to point of purchase, but is sometimes used to refer to the cash register or another place where payment is made.

**POINT OF SALE TERMINAL** A special type of computer terminal which is used to collect and store retail sales data. This terminal may be connected to a bar code reader and it may query a central computer for the current price of that item. It may also contain a device for getting authorizations on credit cards.

**POINT OF TERMINATION (POT)** The point of demarcation within a customer-designated premises at which the telephone company's responsibility for the provision of access service ends.

**POISSON DISTRIBUTION** A mathematical formula named after the French mathematician S.D. Poisson, which indicates the probability of certain events occurring. It is used in traffic engineering to design telephone networks. It is one method of figuring out how many trunks you will need in the future based on measurements of past calls.

Poisson distribution describes how calls react when they encounter blockage (see QUEUING THEORY for a detailed explanation of blockage). There are two main formulas used today in traffic engineering: Erlang B and Poisson. The Erlang B formula assumes all blocked calls are cleared. This means they disappear, never to reappear. The Poisson formula assumes no blocked calls disappear. The user simply redials and redials. If you use the Poisson method of prediction, you will buy more trunks than if you use Erlang B. Poisson typically overestimates the number of trunks you will need, while Erlang B typically underestimates the number of trunks you will need.

There are other more complex but more accurate ways of figuring trunks, most critically Erlang C (blocked calls delayed or queued) and computer simulation. Poisson has been used extensively by AT&T to recommend to its customers the number of trunks they needed. Since AT&T was selling the circuits and preferred its customers to have excellent service, it made sense to use the Poisson formula. As competition in long distance has heated up, as circuits have become more costly and as companies have become more cost-sensitive (more aware of their rising phone bills), Poisson has fallen out of favor.

**POISSON PROCESS** A kind of random process based on simplified mathematical assumptions which makes the development of complex probability functions easier. In traffic theory, the arrival of telephone calls for service is considered a Poisson process. Calls arrive "individually and collectively at random," and the probability of a new call arriving in any time interval is independent of the number of calls already present. A Poisson process should not be confused with the Poisson Distribution, which gives the probability that a certain number of calls will be present if certain additional assumptions are made. See POISSON DISTRIBUTION.

**POP** See POINT OF PRESENCE.

**PORT SHELF** A term for a card cage assembly within an ACD that holds mainly trunk, agent, and device cards.

**POS** See POINT OF SALE.

**POSITION** A telephone console at a switchboard manned, er, staffed by an attendant, or operator, or agent, or whatever the latest fashionable word is.

**POTENTIAL REVENUE** The revenue value per call times the number of calls forecast for a given period.

**POTS** When you hear a telecommunications person talking about POTS, he or she is referring to Plain Old Telephone Service. Just dial tone. No features at all. Not even residential features like call waiting.

**POWER CONDITIONING** Power conditioning is a generic concept that encompasses all the methods of protecting sensitive hardware against electrical power fluctuations. When electricity leaves a commercial power generating plant, it is very clean. Unfortunately, nearly all devices connected to power lines — and the worst are things with motors, like elevators, air conditioners, etc. — create disturbances that pollute the power stream.

As power travels through a wire away from the power plant, it picks up more of these interferences. A pure AC power sine wave appears as a smooth wave. The height of the wave is measured in volts. The wave starts at zero volts and moves to the highest point of 120 volts. The wave then cycles through a low point of -120 volts and back to zero. The speed at which it travels through this cycle is the frequency. Normal frequency in North America is 60 cycles per second (Hz). (In other places it's often 50 cycles per second.) Anything that disrupts this wave can cause hardware or data problems and needs to be regulated.

Power disturbances can be categorized in several ways. A transient, sometimes called a spike or surge, is a very short, but extreme, burst of voltage. Noise or static is a smaller change in voltage. And brownouts and blackouts are the temporary drop in or loss of electrical power. (The terms "brownout" and "blackout" are never used by electrical engineers; they are popularizations that refer to a broad range of electrical conditions.)

Three types of protection against these three events are available: suppression, isolation, and regulation.

Suppression protects against transients. The most common suppression devices are surge protectors that include circuitry to prevent excess voltage. Although manufacturers originally designed surge protectors to prevent large voltage changes, most have also added circuitry to reduce noise on the line. Isolation protects against noise. Ferro-resonant isolation transformers use a transformer within the circuitry to envelop the sine wave at a slightly higher and lower voltage. Any voltage irregularity that extends beyond this envelope is clamped. Isolation transformers are usually expensive.

Regulation protects against brownouts and blackouts. Regulation modifies the power wave to conform to a nearly pure wave form. The Uninterruptible Power Supply (UPS) is the most commonly used form of regulation. A UPS comes in two varieties, on-line and off-line. An on-line UPS actively modifies the power as it moves through the unit. This is closer to true regulation than the off- line variety. If a power outage occurs, the unit is already active and continues to provide power. The on-line UPS is usually more expensive but provides a nearly constant source of energy during power outages. The off-line UPS monitors the AC line. When power drops, the UPS is activated. The drawback to this method is the slight lag before the off-line UPS jumps into action. That lag is getting shorter as electronics improves. So it's rarely a problem any longer.

Because UPS systems are expensive, most companies attach them only to the most critical devices, such as phone systems, network file servers, routers, and hard disk subsystems. Attaching a UPS to a local area network file server enables the server to properly close files and rewrite the system directory to disk. Sadly, most programs run on the workstation and data stored in their RAM is not saved during a power outage unless each workstation has its own UPS. If the UPS doesn't have its own form of surge protection, it is a good idea to install a surge protector to protect the UPS from transients. Proper use of power conditioning devices greatly reduces telephone system and network maintenance costs. Make sure that proper amperage is available for each system and that all outlets are grounded. Power conditioning devices connected to poorly-grounded outlets offer very little protection.

Studies have shown that total local area network maintenance costs are higher with line-surge suppressors and ferro-resonant isolation transformers alone, than with uninterruptible power supplies.

**POWER DIALER** An automatic dialing system that shares some characteristics of predictive dialing and preview dialing, but is actually neither. It consists of a piece of hardware to do the dialing, plus list management software, often a simple database. Like a predictive dialer, a power dialer can screen out busy signals, no-answers, fax machines, answering machines, voice mail, and all the other non-responsive answers. But unlike predictive dialing, there is no algorithm which calculates ahead of the agent and places calls in anticipation of that agent being ready. You can theoretically use this kind of dialing in applications of one or two agents (situations in which predictive dialing is meaningless). Like preview dialing, it may require (or permit, depending on your point of view) the rep to choose the next call.

Another flavor of power dialing is something best described as a repetitive autodialer: a number-crunching system that places a call every X seconds without a specific representative assigned to serve the call. This too screens out no-answers and busies. But a major drawback is that without a predictive algorithm, there might not be anyone available, and the callee hangs up in disgust.

Unfortunately, the term power dialing is sometimes used by autodialer vendors interchangeably with predictive dialers, or to connote something more "powerful" or advanced than regular old outdialers without regard to specific features. The problem is that a dialer has to meet certain criteria to be called either predictive or preview. If a vendor makes a dialer that doesn't have a predictive algorithm, but they want to position it in the marketplace as more advanced than a mere preview dialer, they may tag it as a "power" dialer, confusing all and enlightening none. As the term with the loosest definition, it's rapidly losing any real meaning.

**POWER POLE** A pole used to run cable (electric, computer and telephone) from the ceiling to a desktop that does not have a wall or pillar nearby. Most often used in an "open plan" office (a big empty room filled with desks or carrels), especially call centers or data-entry departments.

**PRE-QUALIFICATION** If a potential customer spends money to respond to a promotion, he or she is considered a pretty good lead. So callers to 900 number promotions are said to be "pre- qualified" because they are interested in the product offered despite the cost of getting involved. They qualify themselves. These are the best customers.

**PREDICTIVE DIALER** See PREDICTIVE DIALING.

**PREDICTIVE DIALING** A method of making outbound calls that uses advanced software to estimate the correct number of calls to place, and the number of agents that will be able to handle those calls.

Predictive dialing automates the entire process, with the computer choosing the person to be called and dialing the number. The call is only passed to the agent when a live human answers. Predictive dialers screen out all the non-productive calls before they reach the agent: all the busy signals, no- answers, answering machines, network messages, and so on. The agent simply moves from one ready call to another, without stopping to dial, listen, or choose the next call.

True predictive dialing is merely one kind of automated dialing — there are others. But predictive is the most powerful and the most productivity-enhancing. True pre-dictive dialing has complex mathematical algorithms that consider, in real time, the number of available telephone lines, the number of available operators, the proba-bility of not reaching the intended party, the time between calls required for maxi-mum operator efficiency, the length of an average conversation and the average length of time the operators need to enter the relevant data. Some predictive dialing systems constantly adjust the dialing rate by monitoring changes in all these factors. The dialer is taking a sort of gamble: knowing that these processes are in motion, and knowing that there is a certain chance that a call placed will end in failure, it must throw more calls into the network than there are agents available to handle them, if they all succeed. Sometimes the prediction is wrong, and there are fewer

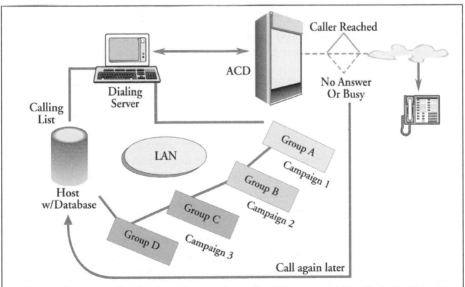

In this example of how predictive dialing software works, a calling list is programmed into the host for three dif-ferent campaigns. The dialing server goes to the database and dials for a group of agents on a LAN, sending the call to the first available agent if it reaches a live voice. If there is no answer or the system gets a busy signal, the dialing system sends the number back into the database to dial again later.

failures than expected. In this case the called party will pick up the phone, say hello, and be hung up on when no agent is available. One of the intricacies of predictive dialer management is finetuning the aggressiveness of your dialer's algorithm.

Predictive dialing has been nothing short of revolutionary in the outbound call center. When operators dial calls manually, the typical talk time is close to 25 minutes per hour. Most of the rest of that time is non-productive: looking up the next number to dial it; dialing the phone; listening to the rings; dealing with the answering machine or the busy signal, etc. Predictive dialing takes all that away from the agent's desk and buries it inside the processor.

When working with a predictive dialer, it is possible to push agent performance into the range of 45 to 50 minutes per hour. We've heard of centers going as high as 54 minutes per hour. (You can't go higher than that, taking into account post-call wrap up time.)

There is more to the technology than just the pacing algorithm. Predictive machines excel at detecting exactly what is on the other end of the phone, including the ability to differentiate a human voice from an answering machine. They typically decide that the call has reached a person within the first 1/50th of a second — the start of the word "hello."

Predictive dialing has always been a software application. It required a great deal of processing power, so the vendors put their specialized software onto high-powered computers, most of them with a closed architecture. But the research and development was always geared to better dialing algorithms, more sophisticated call tracking features, and better database management — essentially software apps.

What started as a great idea for outbound telemarketing — fire out more calls than necessary to maximize agent productivity — became the platform on which software companies continued to refine and develop new features for handling calls.

It was such a good idea that companies in other areas (telemarketing software, especially) began adding predictive dialing modules to their systems. The logic was good: if dialing features are mainly software, and powerful generic processors are available to run them, there's no reason not to create a whole new category of product — the PC-based (or at least client/server-based) dialer.

The traditional hardware/dialing vendors are now changing to match. Several of them have taken their core technologies, enhanced them, and are presenting them to call centers in a new light. They are creating systems for managing all aspects of the call flow. They let agents make calls in predictive mode, and receive incoming calls as well.

They let you connect peripherals like voice response units to their systems. And some, finally, let you develop applications to sit on top.

**PREVIEW DIALING** Preview dialing is a term used to describe an automatic dialer. Preview dialing is also called "screen dialing" or "cursor dialing." Typically the prospect's account information and/or phone number appears on the screen BEFORE the call is made. Thus the agent can "preview" the number, the screen, the customer. If the agent wants to make the call, the agent hits a key and the computer dials the number. The agent will also still hear the mechanics of dialing —the tones, rings and clicks. It's primarily a business-to-business mode.

In some preview dialing equipment, the agent must hit a key if he/she doesn't want the number dialed. Contrast preview dialing with predictive dialing, where the computer makes all the dialing decisions and presents the calls to the agent only after they are connected. Predictive dialing is a lot faster than preview dialing. See PREDICTIVE DIALING.

**PRIMARY AGENT GROUP** An automatic call distributor term: the main set of agents for which inbound calls are intended. You'd use this term to refer to the place where calls are supposed to go. When they can't go there (if all agents are busy), then they pass to secondary and tertiary groups. The decision on when to pass from primary to secondary depends on the overflow parameters you set. Different ACD systems label this process differently.

**PRIMARY AVERAGE POSITIONS MANNED** See PAPM.

**PRIMARY RATE INTERFACE (PRI)** The ISDN equivalent of a T-1 circuit. A phone line that provides 23 bearer channels and one data channel running at 1.544 megabits per second and 2.048 megabits per second, respectively.

**PRIVATE AUTOMATIC BRANCH EXCHANGE** See PABX.

**PRIVATE BRANCH EXCHANGE PBX.** Term used now interchangeably with PABX. PBX is a private telephone switching system, usually located on a customer's premises with an attendant console. See PBX.

**PROBABILITY CODE** Some telephone lookup software will show the user a code that tells him how good a match is, based on the reliability of the sources for the data used to make the match. That code is expressed as a probability of accuracy.

**PROBLEM RESOLUTION ENGINE** The component of help desk software that assists in finding solutions to customer problems. Using input from the customer (usually filtered through the first line agent) it tries to match the problem as described with a database of possible fixes. There are many methods of finding the right answer, none of them perfect, many of them very good.

**PRODUCTIVITY** In general, a word that means how much work gets done in a certain amount of time. Lately it is being used more frequently instead of the term "occupancy." When you think about it, the word "productivity" is much more business-like and bottom-line than "occupancy," but it probably gives a false impression of the relationship between how much work gets done in a certain amount of time and how much time an agent spends talking to callers in a certain amount of time. (If she doesn't take an orders, make any sales or solve any problems, the agent hasn't been very productive, has she?) We vote for keeping "productivity" in its usual English usage and not as a call center term.

**PROGRESSIVE DIALING** A method of dialing multiple numbers that is more automated than preview dialing, but not by much. The customer record is not displayed until the number is dialed, giving the agent less time to review it and a shorter period to rest between calls. Also, the agent does not control the sequence of numbers. See PREVIEW DIALING, PREDICTIVE DIALING.

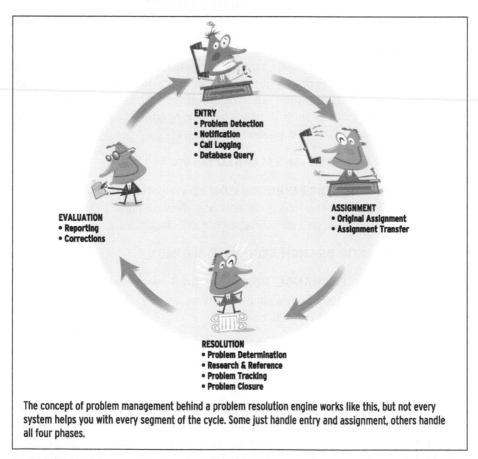

**ENTRY**
• Problem Detection
• Notification
• Call Logging
• Database Query

**ASSIGNMENT**
• Original Assignment
• Assignment Transfer

**RESOLUTION**
• Problem Determination
• Research & Reference
• Problem Tracking
• Problem Closure

**EVALUATION**
• Reporting
• Corrections

The concept of problem management behind a problem resolution engine works like this, but not every system helps you with every segment of the cycle. Some just handle entry and assignment, others handle all four phases.

**PROPRIETARY** If something is proprietary it means it will only work with one vendor's telephone system. There are many telephones that are proprietary to one telephone system or one manufacturer. These proprietary phones are usually the electronic and multi-line instruments. In a more general sense, you will find many kinds of proprietary systems in a call center, from certain computer systems to predictive dialers — proprietary meaning that the manufacturer won't let you hook in anybody else's peripherals.

**PROPRIETARY TELEPHONE SETS** Proprietary telephones are feature phones that are specific to a particular make of PBX, ACD or other switching system. They may be digital or analog. As they are custom-designed for that system, they have non-standard interfaces and have non-standard protocols to communicate between the telephone and the switch. This has several implications:

1. You can't take a proprietary phone from one switch and expect it to run on another switch.

2. Proprietary phones are expensive, and are highly profitable to the their manufacturers. Hence the manufacturers' insistence on keeping them proprietary.

3. Signaling between proprietary phones and their switches is richer than signaling between switches and single line analog phones. As a result, it's preferable to integrate voice mail and automated attendants through proprietary phones. Sometimes the manufacturer of the switch will divulge his secret signaling scheme. Other times he won't.

**PROSODY** Intonation. In text-to-speech, prosody refers to how natural the speech sounds — the ups and downs of the sentence. (English majors will also recognize this term from Poetry 101.)

**PROSPECT** Used as a general term for any person or company that is a potential customer for your product or service. That is, they have contacted you to ask about your product or service or they have shown interest after a cold call.

In some cases the term is used more specifically to refer to a "suspect" (same as the above meaning of prospect) who has passed certain buying criteria set by you or your company. That is, they have enough money to pay for it and may actually be able to use it. In this sense a prospect is a qualified suspect.

**PROTOCOL** A set of rules governing voice or data communications. They set the parameters for the machines at each end of the link to stay connected and monitor such things as file transfers, error correction, and flow control. Protocols can be either proprietary, meaning that only devices made by the same vendor can talk to each other; or standard, meaning that one of the national or international governing bodies has codified the rules for transmission of that type. Fax protocols, for example, are regulated by the International Telecommunications Union (ITU) out of Geneva. So, for that matter, are most telecom protocols. For more information on who promulgates what standards, check out www.AlternateTelephony.com.

**PSEUDOTERNARY** Sounds like some kind of dinosaur, or a least a medicine, but "pseudoternary" is an ISDN term. It's used to describe ISDN basic rate interface data coding. What it means is this: binary ones are represented by no line signal and binary zeros are represented by alternating positive and negative pulses. That's three encoded signal levels representing two-level binary data.

PSTN Public Switched Telephone Network. When you are home and you call your best friend, across town or across the country, you are using the Public Switched Telephone Network or PSTN. The PSTN is the regular old telephone system, which anyone can use by picking up a telephone. In telecom charts and schematics it is represented by a cloud. Without private networks and virtual private networks, there would probably be no use for this term. Private networks, virtual private networks and dedicated access lines are NOT part of the PSTN. Just about everything else is.

**PSYCHOGRAPHICS** Market research based on lifestyle habits and preferences. Psychographics is the study of how to relate habits to purchasing decisions. It looks for relationships, and for the data to be able to make those connections. For example, if the researcher thinks that people who play a lot of squash buy a lot of wine coolers, they will look for lists of people who read squash magazines and who returned warrantee cards for squash racquets. Then those people will get a call asking them to try the wine cooler. The results are then analyzed to see if there is a relationship.

**PUBLIC SWITCHED TELEPHONE NETWORK** See PSTN.

**PUBLIC UTILITIES COMMISSION** See PUC.

**PUC** Public Utilities Commission. The state agency that regulates telephone service within that state. (They also regulate the state's other public utilities like electric power.) Regulation of nationwide telecommunication is handled by the FCC.

**PULSE CODE MODULATION** Pulse Code Modulation. The most common method of encoding an analog voice signal into a digital bit stream. First, the amplitude of the voice conversation is sampled. This sample is then coded into a binary (digital) number. This digital number consists of zeros and ones. The voice signal can then be switched, transmitted and stored digitally. PCM refers to a technique of digitization. It does not refer to a universally accepted standard of digitizing voice. The most common PCM method is to sample a voice conversation at 8000 times a seconds.

**PURGE** To remove duplicate or other unwanted names from a list.

**QUALIFY** When dealing with potential customers, the process of separating those who are likely to buy from those who are not. Essentially qualification involves comparing a person who has shown an interest in your product or service to a list of criteria that establishes who you think your customers are.

Some criteria that are used to separate prospects from suspects (as unqualified prospects are sometimes called) are:

• Are they able to pay for the product or service (a high enough income level or credit background)?

• Is the company big enough or have a high enough volume (or small enough as the case may be) for your product to actually meet their need?

• Are they located in an area where your product is sold, or your service offered?

Separating qualified leads from unqualified ones is important. The cost of selling some products is so high that it becomes like a form of triage, and finding those most likely to buy is the only way to avoid expensive failure. In a call center, the initial information gatherer is the first line agent, who after qualifying the prospect would then assign the call to an appropriate senior seller, or handle the tasks of sending literature to the caller (often through an automated recordkeeping and fulfillment system).

**QUALITATIVE DATA** Market research information gathered in ways that are less rigorously scientific than surveys and other quantitative methods. Focus groups are an example. Companies use them when they need to generate ideas instead of conclusions, says Mark Green of Market Intelligence. "One or two very persuasive people can sway a group," he says "Focus groups have a tendency to create ideas as well as measure them."

**QUALITY OF SERVICE** Sometimes used as a telephone-carrier equivalent to SERVICE LEVEL. To a telephone carrier, it means a measure of the telephone service quality provided to a subscriber. It's not easy to define "quality" of the telephone service. Is the call easy to hear? Is it "clear?" Is it loud enough, etc.? The state public service commissions have various measures which they insist phone companies conform to. They tend to be more measurable. They include the longest time someone should wait after picking up the handset before they receive dial tone (three seconds in most states).

**QUANTITATIVE DATA**  Market research information that takes specific measures in a scientific, organized, repeatable way. Surveys, for example, gather quantitative data about the number of people who answer a specific question a certain way.

**QUEUE**  A stream of tasks waiting to be executed. A series of calls waiting to be answered. In call centers, this is often called the "hold queue." It used to be that a switch would hold the queue, but with the advent of multi-site call centers, with multiple linked switches (often from different vendors). it is now possible to actually queue a call in  the phone network, while the various centers are polled to see where the call should be sent. This is done through a combination of switch technology and software intelligence that's built into the network (and offered as service by the carriers).

**QUEUE MANAGEMENT**  the process by which the switch, or the network, or any decision-making entity, lines up calls (and other "transactions" like faxes and IVR requests) and chooses the order in which those transactions occur. Managing the queue involves a graceful combination of randomness (you never know what the caller will do), prioritization (some callers may be more important than others, and with the proper tools in place you can know who they are). and pre-defined choices (for example, calls for service can stay in queue longer that call for sales, or vice versa).

**QUEUE SERVICE INTERVAL**  The maximum length of time a queue will go unsampled.

**QUEUED MODE**  Calls entering an Automatic Call Distribution system wait in a queue and are presented, one at a time, to the first available agent in the chosen group.

**QUEUING**  The act of "stacking" or holding calls to be handled by a specific person, trunk or trunk group.

**Queuing Theory**  The study of the behavior of a system that uses queuing, such as a telephone system. Much of queuing theory derives from the science of Operations Research (OR). Dr. Leonard Kleinrock has written the authoritative books on the subject. Here is an explanation of Queuing Theory from James Harry Green's Dow Jones-Irwin Handbook of Telecommunications.

"The most common [telephone] network design method involves modeling the [phone] network according to principles of queuing theory, which describes how customers or users behave in a queue. Three variables are considered in network design. The first is the arrival or input process that describes the way users array themselves as they arrive to request service...The second variable is the service process which describes the way users are handled when they are taken from the queue and admitted into the service providing mechanism. The Third method is the queue discipline, which describes the way users behave when they encounter blockage in the network...Three reactions are possible:

• Blocked calls held (BCH). When users encounter blockage, they immediately redial and reenter the queue.

- Blocked calls cleared (BCC). When users encounter blockage, they wait for some time before redialing.

- Blocked calls delayed (BCD). When users encounter blockage. they are placed in a holding circuit until capacity to serve them is available. See QUEUE.)

"Traffic engineers have different formulas or tables to apply, corresponding to the assumption about how users behave when they encounter blockage." See ERLANG, POISSON.

Here is another way of looking at the same phenomenon: Queuing theory can be described as the study of systems in which customers wait in line for service to become available, the "blocked calls delayed" condition in telephony (see TRAFFIC ENGINEERING). Although seldom used in designing voice networks (other techniques are usually more cost-effective), queuing is very important in the design of packet networks where speed of transmission more than offsets the delay of waiting for a transmission facility to become available, and in staffing for Automatic Call Distributors.

**QUICK DISCONNECT** A snap-apart connection the cord between the headset and the phone. It lets you make a quick getaway without taking your headset off. that's important — headsets are prone to breakage, and the less manipulation you do with the headset or the actual phone jack, the longer it will last.

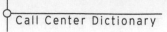

**RAID** Redundant Array of Inexpensive Disks. This is a data protection scheme that involves housing several disk drives in a single chassis, then writing your data over the array in such a way that if you lose one or more of the drives, you still won't lose any of the data.

**RAISED FLOOR** A floor built over the real floor so cable can be run in between. It's the best way to run cable (computer or telephone) to individual desks in an office with few walls (like a call center). It's very expensive, though, so some companies dig a trench into the concrete below the real floor to do the same thing. A raised floor is also sometimes known as a "computer floor."

**RAN** Recorded ANnouncement. What Nortel calls the ACD recorded announcement feature for its Meridian telephone switch.

**RATE CHIP** A standard, nonvolatile memory device used to retain database information on call pricing by Area Code and Central Office. Typically used in call accounting equipment.

**RBOC** Regional Bell Operating Company. Also known as RBHC, Regional Bell Holding Company, or Baby Bell. The old Bell System was broken up into seven RBOCs. Each consisted of one or more Bell Operating Companies (BOCs). For example, Pacific Telesis, an RBOC, consisted of Pacific Bell and Nevada Bell (and several other companies which are not BOCs). Southwestern Bell is the only RBOC that has just one BOC — also called Southwestern Bell. (This was all done by court order, that's why it's weird.) The other RBOCs were: Nynex, Bell Atlantic, Ameritech, US West and Bell South. Of those, just a few remain, due to mergers among them. We're not going to say how many there are right now, because by the time you get this book into your hand, 1999's telecom merger frenzy should be in full swing.

RBOCs and BOCs are regulated by the FCC (for the moment).

**RE-ENGINEERING** A term probably invented by Michael Hammer in the July-August, 1990 issue of Harvard Business Review. In that issue, he wrote "It is time to stop paving the cowpaths. Instead of embedding outdated processes in silicon and software, we should obliterate them and start over. We should 're-engineer' our business: use the power of modern information technology to radically redesign our business processes in order to achieve dramatic improvements in their performance." The term re-engineering now seems to mean taking tasks presently running on mainframes and making them run on file servers running on LANs — Local

**163**

Area Networks. The idea is to save money on hardware and make the information more freely available to more people. More intelligent companies also redesign their organization to use the now, more-freely available information.

The application to call centers is this: call centers literally are how companies re-engineer themselves. They are the perfect example of how to rethink the most basic element of a company's processes: the relationship with the customer. How do you reach the customer, and how does she reach you? What information do you use to interact with that customer? What does that customer expect, and how well do you meet that expectation? What does it cost you to meet that expectation? How do you find new customers? In the modern business, all the answers to these fundamental questions lead through the call center. It's somewhat ironic, then, that call centers didn't figure in most management thinking ten years ago. Nowadays they are considered so critical to business functions that what goes on in them is often a closely guarded secret.

**READERBOARD** A screen that's hung up on the wall in a call center, usually made up of scoreboard-style LED lights, that shows a wide variety of ACD statistics and other informational messages aimed at agents. Readerboards often pull data directly out of the ACD to show people the length of the queue, for example, or the number of calls waiting. Popular models can display information in multiple colors, and can use graphics, so that certain threshold alert conditions can be prominently broadcast throughout a center.

In recent years, readerboards, which are also variously called display boards or visual information systems, have seen competition from both TV monitors hung up on walls, and from Internet-based scrolls that can be pushed across the agent's screen like a mini stock ticker. All these types have advantages and disadvantages. Readerboards are cheap, but you often need several to get them in a place where every agent has a view. Plus, they actually have to look up to see the data. A TV can show you more data, in more different colors and formats, but they are more expensive. And a scrolling or flashing ticker on the screen, now a feature of more different types of agent desktop software than we can describe, can be a distraction, especially if it has enormous potential for customization (which they almost always do). People used to be concerned with the amount of screen real estate they took up, but then monitors got larger, so that's not such a concern anymore.

**REAL-TIME ADHERENCE** Adherence is a term used to connote whether the people working in the center are doing what they're meant to be doing. Are they at work? Are they on break? Are they answering the phone? Are they at lunch? All these activities are scheduled by workforce management software (also called forecasting software, or call center management software).

If they're in line, the workers are "in adherence." If not, they're "out of adherence." Some automatic call distributors have a real-time adherence data link which connects the ACD to an external computer which then tracks and displays current service rep activity measured against a pre-defined schedule. The idea is to give call center supervisors tools to manage the center's workforce more efficiently. Supervisors are able to define the task, the start time of each task, and the task duration. In addition, thresholds and ranges of acceptable deviations for the call center can be set for each task or service rep work state. Once the schedules have been defined and thresholds set, real-time displays inform the supervisor of discrepancies between the work schedule and actual activity. Service rep information will automatically appear should their status exceed the threshold, such as someone being on break for too long.

**REAL-TIME DATA** In a call center reporting system, there are two kinds of information: what happened in the past ("historical" data) and what's happening now ("real-time"). You need them both, for different things. Historical data helps you project future call volume and staffing needs, among other things. Real-time data, by contrast, is used for making instant changes in the call center's operation: moving agents from one group to another, canceling break times if call volume is too high, responding to a n etwork outage or other sudden change.

**REAL-TIME STATUS DISPLAY** In a call center controlled by an ACD, each group of agents is usually monitored by a supervisor. That person's terminal shows the status of each agent (whether he is on a call or available to take one), of each line coming into the group, and of the number of calls waiting to be answered.

There is considerable variation in what vendors will describe as real-time. Many, if not most, good systems will update as quickly as every five seconds. It's fair to say that a terminal's display can be considered in "real-time" if the ACD updates it roughly every 30 seconds. Newer ACDs are starting to show the information graphically, using color, so that the supervisor can see at a glance when conditions change.

**REBILLER** A rebiller, also called a switchless reseller, buys long distance service in bulk from a long distance company, such as AT&T, and resells that service to smaller users. It typically gets its monthly bill on magnetic tape, then rebills the bulk service to its customers. A rebiller owns no communications facilities — switches or transmission. It has two "assets" — a computer program to rebill the tape and sales skills to sell its services to end users. The profit it makes comes from the difference between what it pays the long distance company and what it is able to sell its services at. It's not an easy business to be in, since you are selling a long distance company's services to compete against itself.

**RECENCY** The date of last purchase or activity for a customer on a list. Or, the names of those who have made the most recent purchases on a given list. Along with frequency and monetary value, this is one of the most important things you can know about a name on a direct response list.

**RECORD** An entry in a database. A record is a set of related pieces of information, linked by the format of the database file. For instance, in a database of customers, each record would consist of a name, phone number, and address, each in a separate field.

**RECORD LOCKING** A feature of networking computers that facilitates file sharing. When two or more users are accessing the same database file, the records each one changes are inaccessible to the other. This prevents them from inadvertently corrupting the file and wiping out data, or from reporting inaccurate information that has been updated somewhere else.

For example, imagine making an airline reservation. You call up. You want to change your reservation. While the airline has your record open, your travel agent calls up to change it. You change your reservation. Your travel agent changes it. Which one ends up in the "permanent" record? Confusion reigns. Clearly it makes sense to only allow one person to access one record at once and lock everyone else out.

Record locking is the most common and most sophisticated means for multi-user LAN applications to maintain data integrity. Though it doesn't allow users into the same record at the same time, record locking does allows multiple users to work on the same file simultaneously. So multi-user access is maximized. Contrast with file locking, which only allows a single user to work on a file at a time.

**RECORD ON DEMAND** A feature of a monitoring and/or quality assurance tool that enables the user (usually a supervisor) to start an immediate recording session on an agent, rather than (or in addition to) a regularly scheduled session.

**RECORDER** A device many large phone users use to record conversations with their callers. Recording truck dispatches can help a company gain the upper hand in customer service. Purchasing departments may use the recorder to remind vendors of their promises. The financial department can document money transfer orders and investments. Recorders come in several sizes. There are cassette recorders with standard speed and slow extended play speed. Open or reel-to-reel recorders have features similar to cassette recorders. Cassette recorders may be voice-operated (VOX) or started by a recorder coupler.

Some recorders can search for and recall conversations recorded with an option called "autosearch."

**RECORDER WARNING TONE** A one-half second burst of 1400 Hz applied to a telephone line every 15 seconds to indicate to the called party that the calling party is recording the conversation. This tone is required by law to be generated as an integral part of any recording device used for the purpose and is required to be not under the control of the calling party. The tone is recorded together with the conversation.

**RECYCLE** A term for the rebuilding of a data file from previously dialed numbers in which there was no connect.

**REDUNDANCY** A duplicate component within a system (like an ACD) that shares the load with another component, or takes the load if the primary component fails. Some redundant components activate automatically if the primary fails; some require manual activation.

**REFRESH RATE** 1. The rate are which an entire screenful of information is painted on a monitor from top to bottom. The lower the rate (expressed in cycles per second, or hertz) the more flicker the eye will detect. This is a useful stat when you buy monitors. 2. The rate at which a display of information from an ACD or other processor is updated on someone's screen by software. Agent status statistics, for example, on a supervisor's screen.

**REGIONAL BELL HOLDING COMPANY** Usually abbreviated RBHC. See RBOC.

**REGIONAL BELL OPERATING COMPANY** See RBOC.

**REGRESSION ANALYSIS** Will adding more agents improve your quality of service? One way to find out is through regression analysis. Regression analysis statistically compares the effect of one variable on another with the goal of finding an equation that can predict what will happen to the second variable when the first one changes. In our example, you might study the relationship between fluctuating levels of staffing and your quality of service, and come up with a formula that will show what will happen to your quality of service if you add two more agents (or whatever).

**RELATIONSHIP ROUTING** A concept introduced by automatic call distributor manufacturer Aspect Communications, to have specific callers routed to agents they had previously developed a relationship with.

**RELEASE**
1. A call comes into a switchboard. The operator calls you to tell you it's for you. Then he/she "releases" the call to you. On most switchboards there's a button labelled "RLS." That's the release button. On some phones (not consoles) the release button is the "hang-up" button. Hitting this button means disconnecting the call. Be careful.

2. The ending of an inbound ACD call by hanging up.

3. The feature key on most ACD instruments labelled Release.

4. A term used in the secondary telecom equipment business. The relinquishing of a piece of equipment to a purchaser or user upon fulfillment or anticipated fulfillment of contractual obligations, whether written or oral.

**REMOTE ACCESS** You are not in your office, but you call in and make a change on your telephone system, voice processing system or computer, usually using the touch tone keypad on your telephone to enter the commands. This feature is called remote access. For example, some automated attendant systems let you call the system from home and change the system's greeting message. If there were an earthquake, you could call in (instead of climbing over the rubble) and change the message to: "Due to the earthquake, our office will be closed today..."

**REMOTE AGENT** An agent who works, not in your main center, but at a remote location, a satellite location or at home. Sometimes known as a telecommuter. To truly be a remote agent and not just someone working by themselves or working in a small call center, the remote agent must receive calls distributed by the main call center.

There are four ways to achieve this. First, the calls can be routed in the public switched network (that is, by your long distance, toll-free or local service provider). Second, calls could be sent to a remote cabinet of your ACD system for further processing. (This is used for remote or satellite locations.) Third, you could use a software-based ACD system and some high bandwidth phone lines to create a virtual call center. Finally, you can buy an add-on product that will connect remote agents to your ACD (or other switch) one agent at a time using either two phone lines, one high bandwidth line or an Internet connection.

The system you choose should return the remote agent's call handling information to the main center so the agent, and the calls handled by that agent, can be properly managed. Customer information can reside locally and be updated in batches or can come from the main call center's systems along with the call.

**REMOTE MAINTENANCE SYSTEM** A Rockwell term for a PC-based workstation that allows support personnel to conduct maintenance activities on the switch (in their case a Spectrum) and obtain system performance information. This can be done either locally or from a remote location over a dial-up modem connection.

**REQUEST FOR INFORMATION** See RFI.

**REQUEST FOR QUOTATION** See RFQ.

**RESELLER** See LONG DISTANCE RESELLER.

**RESPONSE 1.** The level of return interest generated by a promotion or campaign. 2. The answer to a question on a survey.

**RESPONSE ANALYSIS** The responses from a promotion or calling campaign expressed as a percentage or a ratio of the total number targeted.

**RESPORG** Responsible Organization. A company (such as a telecommunications carrier) authorized to interact with the toll-free number assignment database.

**REST TIME** A term for the set time before a dialer begins sending calls to a station after the completion of the previous call.

**RESTRICTION** Phone systems can disallow people or extensions from making certain calls. If they're not allowed to make long distance calls, this is called toll restriction. See TOLL RESTRICTION. There are other forms of restriction, like being able to only use the company's internal network.

**RETRIAL** After failing to complete a call, a person tries again. This is called a "retrial." The term is used in traffic engineering. It's critical in figuring needed trunking capacity. See QUEUING THEORY, POISSON and TRAFFIC ENGINEERING.

**REVERSE MATCHING** Attaching the name and address to a phone number. It's a job usually done by a specialized service bureau. Called "reverse" matching because the service bureaus started in business by attaching phone numbers to lists containing names and their addresses. With ANI (Automatic Number Identification), we get the phone numbers of people calling us. But we don't get their names and addresses. We need to get this information for many reasons, the most obvious being that getting this information on-line and fast saves asking the caller for it and typing all the stuff in. That saves as much as 20 seconds per phone call. (Multiplied by thousands of calls, that equals enormous savings on phone costs; far more is saved than is spent on the list matching.) And asking callers fewer questions about repetitive stuff like phone number, address, city, state, zip means less data entry time spent, and more time to explain the specials we're selling today.

In short, the fewer questions we ask, the less we type and the more stuff we can sell. Reverse matching can be done instantly on-line via a direct data hookup to a distant specialized service bureau or it can be done at the end of the month when we receive our 800 phone bill containing the phone numbers of the people who called us that month.

**RFI** Request For Information. Often the first step in the purchasing process for high tech equipment. A request from a vendor for general information about their product (or products) and services.

**RFP** Request For Proposal. This is step three in the purchasing request process. (Unfortunately they don't go in alphabetical order. Step two is the RFQ if you want to be complete.) This is a formal request made to the last few finalists in a vendor competition. It details specific requirements for the system in question and asks the vendors to make a proposal.

The response to the RFP (the pricing and the particulars) is binding on the vendor. The buyer can sign on the dotted line that is sometimes provided and the response becomes a binding contract or they can deal a little more and draw up a separate contract.

**RFQ** Request For Quote. The buyers give the vendors a pretty good idea what they are looking for and request a pretty close estimate of what such a system will cost.

169

**RISER** The passageway for cabling built or installed between floors of an office building.

**RJ-11** See MODULAR JACK.

**RNA** Representative Not Available. An ACD feature that lets agents take themselves out of the pool of agents available to take a call. A rep might hit the RNA key on her station set because she needs to finish up work from the last call, because she needs to go to the bathroom, because her sleeve is caught in the drawer of the filing cabinet or many other reasons. The RNA function gives the agent an out to deal with any of these matters, then come back and answer calls. RNA time is subtracted from the time an rep is logged in to the ACD to give you the occupancy level. See OCCUPANCY.

**ROBBED BIT SIGNALING** This explanation from Gary Maier of Dianatel: ISDN is the key to future sophisticated telephone network services with its dynamic, highly configurable T-1 connection (also called PRI connection). Since T-1 is a common method of carrying 24 telephone circuits, many wonder about the uses for ISDN, especially when they learn ISDN signaling requires an entire voice channel, reducing today's T-1 from 24 voice channels to 23. But the popular signaling mechanism of "robbed bit" signaling in T-1 has serious limitations. Robbed bit signaling typically uses bits known as the A and B bits. These bits are sent by each side of a T-1 termination and are buried in the voice data of each voice channel in the T-1 circuit. Hence the term "robbed bit" as the bits are stolen from the voice data. Since the bits are stolen so infrequently, the voice quality is not compromised by much. But the available signaling combinations are limited to ringing, hang up, wink, and pulse digit dialing. In fact, the limitations are obvious when one recognizes DNIS and ANI information are sent as DTMF tones.

This introduces a problem: time. Each DTMF tone requires at least 100 milliseconds to send, which in a DNIS and ANI situation with 20 DTMFs will take at least two full seconds. There is also a margin for error in transmission or detection, resulting in DNIS or ANI failures. With the explosion of telephone-related services, the telephone companies are turning to ISDN PRI to provide the more complicated and exact signaling required for new services. ISDN employs a more robust method of signaling. ISDN uses a T-1 circuit as 23 voice channels and one signaling channel. The term 23B plus D refers to 23 bearer (voice) channels and 1 Data (signaling) channel. The data channel carries the signaling information at a rate of 64 kilobits per second. This speed is many times greater than some of the most powerful modems available. Because of this high speed, telephone calls can be placed more quickly, and because of the protocol used, DNIS or ANI transmission failures are impossible.

Additionally, since no bits are "robbed" from the voice channels, the voice quality is better than that of robbed bit signaling on today's T-1 circuits. Also, computer modems and high speed faxes can use the voice channel for sending digital data instead of the traditional analog bit "noise." Therefore, ISDN PRI offers the end user countless new service capabilities. One channel could be used for faxing, another for modem data, several for video, another for a LAN and the remainder for voice. Suddenly, the average T-1 circuit becomes a pipeline for all communications! Increasingly long distance carriers are using ISDN PRI to provide inbound 800 calls with ANI and DNIS and re-routing skills.

**ROTARY DIAL** A circular telephone dial that generates a series of clicks or pulses to represent the number dialed. Dial the number "9" and you will hear nine clicks. People with rotary dial telephones are helpless when they encounter automated attendants, voice mail or other devices that respond to touch tone input. Increasingly, developers are working to address the needs of these callers, including options in those systems that allow rotary callers to either record the input as a message, or interact with the system through speech recognition. Rotary phones are still present in as many as a quarter of all homes in some areas. (Though it's a much smaller percentage of actual phone sets in use, because of business phones.)

**ROUND ROBIN** This is a method of distributing incoming calls to a bunch of people. This method selects the next agent on the list following the agent that received the last call. See also TOP DOWN and LONGEST AVAILABLE.

**ROUTE** The path that a message takes.

**ROUTE ADVANCE** This feature routes outgoing calls over alternate long distance lines when the first choice trunk group is busy. The phone user selects the first choice route by dialing the corresponding access code. The phone equipment automatically advances to alternate trunks and trunk groups, based on the user's class of service. Route advance is a more primitive form of least cost routing. See LEAST COST ROUTING.

**ROUTE OPTIMIZATION** The process of controlling long distance costs Least Cost Routing, Queuing, Toll Restriction and the use of alternate long distance carriers. See also LEAST COST ROUTING, QUEUING and TOLL RESTRICTION.

**ROUTING TABLE** 1. Incoming Phone Calls: A routing table is a user-definable list of steps which are instructions dealing with an incoming call. Ideally these steps should be addressed and the call treatment begun before the call is answered. A routing table should consist of a minimum of steps that include agent groups, voice response devices, announcements (delay and informational) music on hold, intraflow and interflow steps and route dialing (machine-based call forwarding).

A significant issue in the structure of routing tables is "look-back" capability, where no single previously interrogated resource is abandoned by the system (i.e. an agent group is now ignored, even though an agent is now available, because the ACD does not consider previous steps in the routing table).

2. Outgoing Phone Calls: For a specific calling site, this table lists the long distance routing choices for each location to be dialed. There may be only one choice (route) listed for some or all destinations or there may be several choices for some destinations. (It depends how many outgoing lines and how many outgoing trunk groups you have.) If there are several choices then they will be ranked by some criteria (least cost, best quality, etc.).

3. In data communications, a routing table is a table in a router or some other internetworking device that keeps track of routes (and, in some cases, metrics associated with those routes) to particular network destinations.

**RS-232 INTERFACE** A set of standards that specify the electrical characteristics of a particular type of input/output connection in computers. This connection is most commonly known as a "serial port." It comes in 9-pin and 25-pin types, which are wired differently but compatible if you use a very simple adapter. The RS-232 (or RS-232C) is used for computer/modem connections, or sometimes for direct links between two computers (with a special cord called a "null modem" cable).

**RSL** Request and Status Links. A generic term for linking computers and PBXs. Every manufacturer of phone systems is evolving towards open architecture and their own "RSL." The term RSL, which is too passive, is being replaced with PHI (PBX Host Interface), a term coined by Probe Research. For more information, see OAI. S.100 (ECTF) API specification for developers writing CTI applications in C for a client server environment. Created to simplify the development of telephony media applications (fax, IVR, ASR, etc.) and to make them independent of the server platform and hardware. S.100 allows programmers to find and reserve resources, group and connect them together, dispatch commands to them, manage communication among them and funnel events back to the applications. S.100 also defines a system service called the S ystem Call Router (SCR) which provides a simplified call control model for media processing applications that want to delegate line management and call progress to the S.100 framework. For more complex call control applications, the SCR can be used on top of existing call control APIs like TAPI and TSAPI.

**S.200 (ECTF)** S.200 defines a client server protocol corresponding to the S.100 APIs. It defines the messages that are exchanged between the client application and the resource server. S.200 will enable the mixing of applications and servers from different vendors.

**S.300 (ECTF)** S.300 makes it possible for developers to easily add different vendors' technologies to a computer telephony server without the need for rewriting applications.

**S.900 (ECTF)** Simplifies administration and maintenance tasks in client server computer telephony environments.

**SALES AUTOMATION** See SALES FORCE AUTOMATION.

**SALES AUTOMATION SOFTWARE** A program that allows for rapid and orderly maintenance of contact records by salespeople either at the office or in the field. It should also allow them to send follow-up literature, schedule calls and letters, and view a customer's history at a moment's notice. There are many configurations of sales automation software, ranging from a simple record-keeping program on a PC to a complex, multi-user database that hooks together LANs and laptops. They all benefit the user with faster access to more complete information about potential customers.

**SALES FORCE AUTOMATION** The use of computers and computer software by salespeople to boost their sales. There are two types of sales force automation — those totally self-contained on the computers of salespeople (mostly laptops) and those that communicate with databases and local area networks back at the office over phone lines.

There are many purposes of the phone communication — sending orders in, finding out about back orders, getting updates on "specials," dropping letters and memos in, getting new prices, new products, new technical specs, etc. Salespeople routinely show 10% to 20% sales gains armed with a laptop PC and sales automation software.

**SAPM** Secondary average positions manned. A call center statistic relating to the number of positions assigned as back-up or overflow during a defined time for a particular group.

**SAS** See SALES AUTOMATION SOFTWARE.

**SALTING** Placing "dummy," "seed" or "decoy" names in a list. Lets the list's owner know when and how the list was used and flags unauthorized use. Also known as a seeded list.

**SATELLITE OPERATION** A configuration of multiple ACDs or one big ACD and several smaller ACDs. The configuration gives a company with several locations a unified system of centralized trunks, centralized attendants, overall call detail recording and many of the advantages of a private network.

Sometimes this is done by locating what's called a "shelf" of the ACD remotely — essentially a high-speed link to a live peripheral that both sides see as one unit. Another option is to link actual freestanding ACDs together.

One huge call center we know of operates with a dozen or so satellite centers spun off from one main control center. There are no agents in the main center at all, but 4,500 spread out among the rest of the centers. Those satellites are all over the country, linked by high-speed voice and data lines — almost entirely ISDN.

The key advantage of satellite operation is that one big centralized telephone system can contain most of the intelligence and computer smarts for the total system.

**SCAI** Switch to Computer Applications Interface. A protocol that defines how switches talk to outboard computers, i.e. computers which are external to the switch and contain such a database of customer buying information. Using SCAI, calls and data screens about a calling customer can be presented to the agent simultaneously. See OPEN APPLICATION INTERFACE.

**SCC** Specialized Common Carrier. Another term for a long distance carrier in competition with AT&T. The word "Specialized" came about because these long distance carriers purported to provide "specialized" circuits for business customers. At one stage they were also known as OCCs, or Other Common Carriers (i.e. other than AT&T). These days, both terms are pretty irrelevant, but you may see them in old technical manuals. All long distance carriers — including AT&T — are called IntereXchange Carriers (IXCs).

**SCHEDULE** A record that specifies when an employee is supposed to be on duty to handle calls. The complete definition of a schedule is the days of week worked, start time, break times and durations (as well as paid/unpaid status), and stop time.

**SCHEDULE EXCEPTION** A specific date and period when an employee cannot handle calls or is engaged in some kind of special activity. An absence, meeting, or other work assignment creates an "exception" to the employee's daily work file schedule.

**SCHEDULE INFLEXIBILITY** A phenomenon that tends to create overstaffing in some periods when full coverage is the objective in creating a set of schedules. This is caused by the fact that it is impractical to have extremely short schedules for covering momentary peaks in call volume. To achieve a near perfect match of staff and workload at all times would require shifts of virtually every length; for example, 2-hour shifts, 45-minute shifts, even 15-minute shifts.

**SCHEDULE PREFERENCE** A description of the days and hours that an employee would like to work, used by the automatic assignment process to match the employee to a suitable schedule. In some call centers, each employee can have as many as 10 schedules preferences ordered by priority.

**SCHEDULE TEST** A variation of the scheduling process that allows you to forecast the service quality that will result from using an existing set of schedules.

**SCHEDULE TRADE** A situation in which two employees have agreed to work each other's schedules, or an employee has agreed to work the other's schedule, on a specific date or dates.

**SCHEDULING** Making the timetable of agent hours and shifts for your call center. Takes into account vacation days, breaks, training time, lengths of shifts and forecasting information. A call center software management package (also known as workforce management software) helps you do this.

**SCHEDULING SOFTWARE** Often used to mean the same thing as "call center management software," scheduling software is actually one function in a complete call center management software package. Scheduling software uses historical records of past call traffic to create a staff schedule for some future day, week or month. It assumes that similar periods will have similar needs. For example, it assumes that this Christmas Eve will have similar traffic to last Christmas Eve, that this Tuesday at noon will be like all the previous Tuesdays at noon, and so on.

A good system will not only tell you how many agents you need at a given day and time, but given information like union seniority, shift preferences and vacations, can actually crank out a schedule complete with agent names. Some companies make scheduling software that just happens to be popular in call centers. Other companies include a scheduling package with their call center-specific management software.

**SCP** 1. Service Control Point. Also called Signal Control Point. A database maintained by a local telephone company to store customer records and call processing instructions for toll-free (800 and 888) numbers. SCPs contain the intelligence to screen the full ten digits of a toll-free number and route calls to the appropriate, customer-designated long distance carrier. The SCP supplies the translation and routing data needed to deliver advanced network services. It is separated from the actual switch, making it easier to introduce new services on the network.

**SCREEN MONITORING** Supervisor observation of an agent's activity. In general, monitoring is only audio — the call is recorded for later playback and analysis. Sometimes that audio monitoring happens in real time, with the supervisor listening in while the call occurs. When that happens, screen monitoring allows the supervisor to see what the agent is doing along with what he or she is saying — and if necessary, allowing the supervisor to jump in with appropriate assistance. For an explanation of how this works, see SCREEN SCRAPE.

**SCREEN POP** Screen pop is the act of presenting both a voice phone call and a screen full of customer information simultaneously to the agent. The information to make this possible can come from an IVR system, from ANI or DNIS, or from some other caller input like speech recognition. Screen pop is considered the most basic application of computer telephony — the simplest thing you can do to combine voice and data at the desktop.

**SCREEN SCRAPE** The opposite, almost of screen pop. Traditionally, when you monitor calls in a call center, you take an audio recording and store it for later use, either in training, or for secure archiving. Screen scrape is when you capture either an entire agent transactions (including mouse clicks and entered data) or selected snapshots of the screen, along with the audio, so you can have a better picture of what happened during a call. Yes, today's quality assurance software systems can do this, with varying levels of success.

**SCREEN SYNCH** A colloquial term for sending an telephone service agent a phone call together with a screen of information about the incoming call, e.g. the customer's purchasing record or experiences with your product (if you're a help desk, for example.)

**SCSA** Signal Computing System Architecture. SCSA is a comprehensive architecture that describes how both hardware and software building blocks work together. It focuses on "Signal Computing" devices, which refer to any devices that are required to transmit information over the telephone network.

Information can be transmitted through modems, fax, voice or even video. SCSA defines how all these devices work together. Signal computing systems combine three major elements for call processing: Network interfaces, for the input and output of signals transmitted and switched in telecommunications networks; Digital signal processors and software algorithms transform the signals through low-level manipulation; and application programs provide computer control of the processed signals to bring value to the end user.

SCSA is the common set of standards that telecommunication system manufacturers and computing system manufacturers can use to create computer telephony systems. Dialogic Corporation of Parsippany, NJ announced SCSA in the Spring of 1993.

**SCSA CALL ROUTER** An SCSA definition. A system service of SCSA which provides the basic necessities of inbound and outbound call processing and call sharing to client applications, without those applications needing to be aware of the underlying telephony interface operations.

**SCSA COMPATIBLE** Able to function in an SCSA environment in its native mode.

**SDMA** Station Detail Message Accounting. See CALL ACCOUNTING.

**SECONDARY AGENT GROUP** See PRIMARY AGENT GROUP.

**SECONDARY AVERAGE POSITIONS MANNED** See SAPM.

**SECONDARY EQUIPMENT** Used telecommunications and computer equipment.

**SECONDARY MARKET** The market for used business telecommunications and computer equipment.

**SECONDARY NUMBER OF CALLS HANDLED** See SNCH.

**SEEDED LIST** See SALTING.

**SEGMENT** The part of a list that meets certain criteria. For example, only the women from a list of recent car buyers.

**SEIZED** The state of a trunk or circuit that results when a user or device accesses it, making it busy (unavailable to others).

**SELECTS** Names on a list that meet a desired criteria. These criteria could be gender, monetary value of last purchase, frequency of order or just about anything the list has information about.

**SELECTION CHARGE** Extra charge for a list segment that gives you only names that meet your specific criteria.

**SELF-SUPPORT** The call center equivalent of the gas station self-serve island. Using technology (like IVR, fax-on-demand and the internet) to let customers find their own solutions to problems. The advantage: it's cheap, and available 24 hours a day. The disadvantage: customers might not find the answer.

**SENIORITY** A field in each employee record establishing that employee's seniority for the purpose of automatic schedule assignment. This is typically the employee's hire date (year, month, and day) plus a three-digit "tie breaker."

**SEQUENCE** A pattern of days on and days off as defined in either a schedule preference or shift definition.

177

**SERVER** A server is a shared computer on local area network that can be as simple as a regular PC set aside to handle print requests to a single printer. Or, more usually, it is the fastest and brawniest PC around. It may be used as a repository and distributor of data. It may also be the gatekeeper controlling access to voice mail, electronic-mail or fax services.

These days networks have multiple servers. Servers these days have multiple brains, large arrays of big disk drives (often in redundant arrays) and other powerful features. A $35,000 superserver today can match the performance of a $2 million mainframe of ten years ago. Then again, according to Peter Lewis of the New York Times, the lowliest client today has more computing power than was available to the entire Allied Army in World War II. And you could guide Apollo 11 to the moon with the laptop this book is being written on.

**SERVER-BASED ACD** An automatic call distributor that is built on an open platform. Usually this platform is a PC, but it doesn't have to be. Server-based ACDs have many advantages, the biggest of which is their ability to integrate closely with data processing systems, including your IVR system, your computer telephony application and multimedia applications such as e-mail, fax, Internet and video applications. For example, your server-based ACD can route e-mail or a video call just as well as a voice call. It can mix information from your databases with the incoming call with greater ease than a "closed" ACD could. This data can help route calls or can be provided to the agent along with the call.

There are two main drawbacks to the server-based ACD. First is the size of the platform and the required processing power needed for an ACD. Right now most communications servers, including server-based ACDs, are built on a PC platform, or maybe two PC platforms linked together. ACD applications are processing power hogs and a PC platform will only work for the smallest systems. If you have 100 or more agents, let alone a large call center, you're not going to find a server-based ACD that has been built with your needs in mind.

The second drawback is reliability. Most computers are not very reliable by telecommunications standards. No computer works flawlessly even 99% of the time. There are processing delays and other small glitches that you don't even notice. The standards for telecommunications systems are much higher. One of those unnoticeable glitches could hang up on one of your most valued customers. A crash would shut down your whole system until it's fixed. Few, if any, call centers can tolerate being out of business that frequently for that long.

These two problems can be solved, and vendors have already made progress on the second. They offer redundant and fault-tolerant systems that are much more reliable than the average PC. As demand from larger call center grows, the type of platforms used in server-based ACDs may expand too.

**SERVICE BUREAU** A company that does high-volume marketing for others. Some of the services that can be done are: outbound telemarketing, scripting, setting up and receiving inbound calls (800 or 900), creating a phone program, managing lists of names and numbers, fulfillment, demographic analysis, and direct mail, fundraising, disaster planning and whatever else you don't want to do you yourself in your own center.

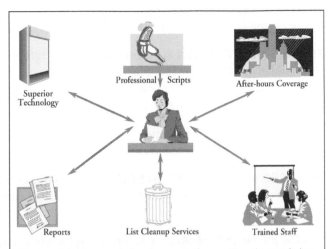

There are lots of reasons to turn to a service bureau. You get access to top-notch technology. (The cost is shared by all their clients.) You get a trained staff, experienced script crafters, reports and convenience. You also get the luxury of testing new ideas without turning your own center upside down.

**SERVICE LEVEL** Usually expressed as a percentage of a statistical goal. For example, if your goal is an average speed of answer of 100 seconds or less, and 80% of your calls are answered in 100 seconds or less, then your service level is 80%. Call centers in different industries have vastly different criteria for measuring successful service. Clearly, a catalog retailer has a vastly lower stake in the outcome of any one call than a cruise line does. Each industry builds their own call center metrics to reflect that.

**SERVICE LEVEL AGREEMENT** A service level agreement is a document that specifies the protocols and procedures for service assurance. It spells out the specific responsibilities of the service provider and the service receiver, for example, who will be notified in the case of particular events, and how that notification will take place. It also spells out what steps will be taken in those cases, and what information will be passed (and to whom) to solve problems.

More detailed SLAs can sometimes explicitly state the performance measures that determine the success or failure of a Service Plan, and go into details on the procedures for problem escalation, enforcement of practices, benchmarks, and (as in any other contract between parties) for renegotiation or amendment. And it may also include the costs for support, specs on hardware and software covered by the agreement, closeout procedures for ending it, etc.

**SERVICE OBJECTIVES** Service objectives define the functional and performance goals for how your system will work and what will be experienced by the system's users and other systems. For example, a switch might have a service objective of answering all incoming calls within ten seconds. Or, a VRU may have a service objective of responding to all user DTMF inputs in less than one second. When we buy or design systems, we often have many service objectives, rarely just one.

**SERVICE OBSERVATION** As a feature of some telephone systems, the Service Observation (SOB) command provides the ability to automatically record data about completed calls, incomplete calls and abnormal calls for the purpose of qualitative supervision of call traffic conditions.

**SERVICE QUALITY** A measure of how well staffing matches workload, expressed often as average delay (in answering a call).

**SERVICES MANAGEMENT SYSTEM** SMS. Administers 800 Data Base Service numbers on a national basis. Customer records for toll-free service are entered into the SCP through this system.

**SEVEN BY TWENTY-FOUR** Open seven days a week, twenty-four hours a day. A way of phrasing a call center's hours of operation.

**SFA** Sales force automation.

**SHIFT DEFINITIONS** A template from which the program can create schedules during a scheduling run. Each shift definition is a record in a Scenario giving more or less precise instructions on shift length, time of day, breaks, and how extensively such schedules can be used in meeting staffing requirements.

**SHRINKAGE PERCENTAGES** A group of Scenario budget assumptions that define the percentage of the time employees are scheduled to work but are not available to handle calls because of absence, breaks, vacation, non-productivity, training, and other activities.

**SHRINK-WRAPPED CALL CENTER** A somewhat wistful term used to describe one possible call center of the future, where all the components of the center (call management, IVR, fax, automated attendant, sales for automation and reporting) are available in one genuinely turnkey product. In another possible future, those functions will be completely outsourced to a service bureau that will handle every aspect of telecom for businesses, including call center functions.

Similarly, any "shrink-wrapped" system, whether it is software or another call center technology is sold as-is. In theory, the product is ready to be used the moment you buy it, but in practice, you have to do any customization (which may be minor) yourself. Think of the —literally — shrink-wrapped software you buy at Staples and you'll get the idea.

**SIGN ON/SIGN OFF** The process of identifying oneself to a machine so as to gain access. In the case of an ACD system this process allows statistics to be kept for this person individually. It also allows for the movement of the person around the system while statistics are accumulated in one logical file.

**SIGNALING SYSTEM 7** SS7. The collection of software systems that drive the public long distance network, and endow it with the ability to carry calls. It provides the hardware with the ability to manage the setup, routing, and information-passing that is at the heart of every phone call. There are three components to signalling: supervising, or watching the line to see its status; alerting, or indicating the arrival of a call; and addressing, or sending the routing information about a call over the network.

SS7 is a powerful tool for moving incredible volumes of calls around the network. It provides fast call setup, through high-speed circuit-switched connections. And it has "transaction capabilities which deal with remote data base interactions," in the words of one expert. What this means is that Signaling System 7 information can tell the called party who's calling and, more important, tell the called party's computer. Signaling System 7 is an integral part of ISDN. It lets us extend services like call forwarding, call waiting, call screening, call transfer, and so on, outside the switch to the full international network. In effect, with Signaling System 7, the entire network will acquire the "smarts" of today's smartest electronic digital phone switch.

**SIGNIFICANT HOUR** Any hour that influences the sizing of a trunk group.

**SILENT MONITORING** Simply put, silent monitoring is when an agent in a call center is being observed (or listened in on) without being aware of it. This is actually typical of monitoring, which is usually consensual — the agent knows it happens in the center, just doesn't know which particular calls will be observed.

**SINGLE DAY ASSIGNMENT** A method of automatic assignment (of employees to Master File schedules) that operates only on schedules of exactly one workday in length. Unlike multi-day assignment (which operates on schedules of any length), this makes it possible for each assigned employee to be scheduled for different hours each day.

**SITERP** SiteRP stands for Sprint Interface to External Routing Processor. It is a service which Sprint offers to enable customer premise equipment (e.g. ACDs) to interact with its long distance network to give that network information on how to route incoming toll-free calls.

With SiteRP, a call processing query is sent to the Sprint SCP (Signal Control Point). The SCP is instructed to check a user database kept on the customer premises for instructions on how to handle the call. Once those call handling instructions have been extracted from the user database, the customer equipment sends a message back to the Sprint SCP, which, in turn, uses that information to decide which call processing instructions it will pass back to the network switch.

SiteRP allows such services as real-time switching of inbound calls, to allow for load balancing among agents or for specific calls to go to specific agents, etc. There are a million reasons why inbound calls from different places and different numbers should go to different agents. To use SiteRP, users must directly connect their SiteRP processors, which can be a mainframe or a PC (or in between) to each of Sprint's five SCPs via 56 Kbps dedicated links.

**SITE-SELECTION CRITERIA** The factors that go into deciding where to put your call center. They include criteria like the cost of labor and land, the educational level of the local workforce, the availability of telecom services and lines, the tax climate, and sometimes the local accent.

**SIX DIGIT TRANSLATION** The process of examining a dialed number — by looking at its first six digits including the area code — to figure out what's the cheapest way to send that call.

Six digit translation is often an integral part of Least Cost Routing programs within the phone system. There are typically two types of "least cost routing" translation — one that examines the first three digits of the phone number (i.e. just the area code) and one that examines the first six digits of the phone number (i.e. the area code and the three digits of the local central office). Six digit translation is preferred because it allows you more flexibility in routing, particularly to big area codes, like 213 in LA, where there are long distance calls within the area code.

**SKILL GROUP** In an ACD (or other call routing device), an agent group that's made up of reps who are qualified to accept calls because of some ability defined in the system. This could be the ability to speak a second language, or a qualification to handle a particular type of customer (priority customers, for example, or support calls). It is interesting to note that this group does not always have to be physically located in the same space. With many switches, they can be spread out across a call center o r a network of centers.

**SKILL MAPPING** An Aspect term for software that enables call center managers to evenly distribute calls to agents based on each agent's special skills. See SKILLS-BASED ROUTING.

**SKILLS-BASED ROUTING** A method of routing incoming calls based on matching the type of call with the defined skills of agents. In other words, someone calling about a broken refrigerator should be directed to a refrigerator expert, not a vacuum cleaner expert. Someone who speaks Spanish goes to a Spanish-speaking agent. The process also covers defining overlapping skills and call types. So that if a refrigerator call comes in from a Spanish-speaking person, you have already prioritized the kinds of skills to make available for that call.

**SKIP TRACING** In collections, the labor-intensive and often tedious process of finding debtors who have moved ("skipped"). Collectors can use many of the same techniques marketers do to search databases for people who have recently moved and to attach phone numbers to names and addresses. And then they can use tools like predictive dialers to speed up the process of calling those people (Call center automation has greatly changed the business of collections in the last few years.)

**SMART HUNT GROUPS** What you get when you implement skills-based routing in your ACD. Once you identify all the unique pieces of knowledge that affect calls ("skills" like the ability to speak a foreign language, familiarity with one or more products, experience with a particular type of call), your ACD sees each call as falling into one or more of the buckets available based on the skills in the center at that particular moment. The process of moving the call into the right bucket is called skill-based routing. The group of people capable of handling a particular call is the smart hunt group. Note that the smart hunt group is not necessarily the same as your defined agent groups — because skills may cross your traditional group boundaries.

At least one ACD vendor has dispensed with traditional ACD groups altogether. Siemens' ResumeRouting feature considers the skills of everyone in the entire call center as a whole for each and every call that comes in. Essentially, you have a collection of micro-groups created on the fly each time you get a call. Each agent has a "resume" of skills used to assess their fitness to handle the call. Good idea because it forces you to assess agent skills, and makes it easy to see who's working out and who's not. The term "smart hunt group" was coined by Rick Luhmann, editor of Computer Telephony Magazine.

**SMDA** Station Message Detail Accounting. Another name for telephone call accounting. See CALL ACCOUNTING SYSTEM.

**SMDR** Station Message Detail Reporting. The basic informatin required for call accounting. See SMDR PORT and CALL ACCOUNTING SYSTEM.

**SMDR PORT** Modern switches have a Station Message Detail Recording (SMDR) electrical plug, usually an RS-232 port, into which one plugs a printer or a call accounting system. The telephone system sends information on each call made from the system to the outside world through the SMDR port. That information — who made the call, where it went, what time of day, etc. — will be printed by the printer or will be "captured" by the call accounting system on a floppy or a hard magnetic disk and later processed into meaningful management reports. See CALL ACCOUNTING SYSTEM.

**SMILE AND DIAL** A condescending term for a no-frills outbound telemarketing operation or technique. Usually heard in the negative: "We train our agents extensively. They are experienced and knowledgeable. We are not doing smile and dial here."

**SMS/800** Services Management System for 800 numbers. The central, national database for 800 number assignments and call processing instructions. It is maintained by DSMI, a subsidiary of BellCore, with day-to-day operations handled by a subsidiary of Lockheed. See DSMI.

**SNCH** Secondary number of calls handled. An ACD statistic. The number of calls handled by a position that were intended for a different ACD group. The number of calls handled by a position as a backup to another group.

**SO** Serving Office. Central office where IXC (IntereXchange Carrier) has POP (Point Of Presence).

**SOFT DOLLAR SAVINGS** Intangible savings realized from the purchase of a product. See HARD DOLLAR SAVINGS.

**SOFT KEY** There are three types of keys on a telephone: hard, programmable and soft. Hard keys are those which do one thing and one thing only, e.g. the touchtone buttons 1, 2, 3, and so on. Programmable keys are those which you can program to do produce a bunch of tones. Those tones might be "dial mother." They might be "transfer this call to my home for the evening." They might be "go into data mode, dial my distant computer, log in and put in my password." SOFT keys are the most interesting. They are unmarked buttons which sit below or above on the side of a screen. They derive their meaning from what's presently on the screen. And what's on the screen will change based on where the call is at that moment — in a conference call, about to set up a conference call, about to go into voice mail, into voice mail, programming a speed dial number, etc.

**SOLID STATE TRANSFER** A power protection term. Some UPSes use "mechanical relays" to switch from AC power to battery power. This technology requires 12 milliseconds of switching time. That's slow enough to cause data loss. Solid state switching is faster and eliminates this kind of malfunction.

**SONET** Synchronous Optical NETwork. A family of fiber-optic transmission rates from 51.84 Mbps to 13.22 Gbps, created to provide the flexibility needed to transport many digital signals with different capacities, and to provide a standard for manufacturers to design from.

**SOURCE/DESTINATION ROUTING** A term for routing calls based on where they originate or terminate.

**SOURCE CODE** The brief code in a list entry that tells you where that name originated. Names from your house list might have "HSE" or "1" on the top line or next to the name. Names you bought from a trade magazine might have "CCM" or "324," and so on. Not to be confused with the computer programming term for the original, alterable version of a software program.

**SORT SCHEME** A list of fields that tells a database program how to sort a report or a list of records. This can be simple scheme that sorts by only one field or a complex scheme consisting of sorts within sorts.

**SPAN**
1. Refers to that portion of a high speed digital system than connects a CO (Central Office) to another CO or terminal office to terminal office.

2. Also called a T-Span Line. A repeatered outside plant four-wire, two twisted-pair transmission line.

3. The total duration of a schedule from start time to stop time, including all breaks.

**SPAN LINE** A T-1 link.

**SPEAKER DEPENDENT** A type of speech recognition which "learns" to recognize a user's voice. The user reads words or phrases that will be used by the system or application several times while the system records and analyzes the similarities and variations.

The disadvantage is that only people who have "trained" the system (by reading in their samples) can use it. The advantages are it can be used as a security device and that people with accents or speech impediments can use it without difficulty.

**SPEAKER INDEPENDENT** A type of speech recognition that can be used with any speaker - one that doesn't have to be trained to understand the voice of a particular person. This is the dominant type of speech recognition used in telephone and call center applications, because it's more useful in situations where a large number of people will interact with it. These types of systems almost always have smaller vocabularies than speaker dependent speech recognition. See SPEAKER DEPENDENT and SPEECH RECOGNITION.

**SPEAKER RECOGNITION** Having a machine recognize human voice. This is an imprecise term. See SPEECH RECOGNITION.

**SPECTRUM** An ACD manufactured by Rockwell International.

**SPEECH CONCATENATION** A term used in voice processing for the kind of recorded speech that uses discrete pre-recorded pieces strung together in a passing resemblance to natural-sounding language.

For example, order status, bank balances, bus schedules or lottery results. Concatenation is done for speed and economy. It lends itself to limited and structured vocabularies that are best stored in RAM (Random Access Memory) or speedily accessible from disk. Concatenation does not replace Text-To-Speech (TTS) as a method of getting the voice processor to deliver its responses. Concatenation, however, can be an excellent complement to TTS when a voice application demands broad, real time vocabulary production. See TEXT-TO-SPEECH.

**SPEECH RECOGNITION** A technology that lets a machine understand a human voice. It allows input into a computer or telephone system through spoken words. A common example is a cellular telephone that can be dialed by speaking the digits of the number to be dialed. A familiar, but completely fictional, example is the way characters on all the Star Trek television shows and movies query their computers. Don't confuse this with the computer or telephone giving information through voice or speech. That's called text-to-speech and is a completely unrelated technology. Some people use the term "voice recognition" to mean the same thing as "speech recognition," but the industry itself prefers speech recognition. (There was too much confusion with voice response the other way.)

There are two kinds of speech recognition. One works with only one speaker who has trained the system (speaker dependent), the other works with any person who uses the system (speaker independent). Speaker dependent speech recognition systems can understand many more words accurately than a speaker independent system can. A speaker independent system has to be able to decipher accents, speech impediments and personal quirks — all of which require more programming and more processing power. Interference from use over telephone lines always taxes a speech recognition system.

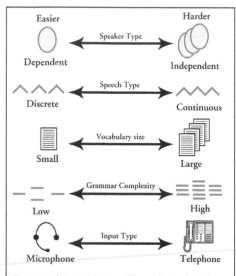

How accurate a speech recognition system is depends on several factors. Performing speech recognition over the phone automatically makes it more difficult. Other factors include the number of words recognized, whether speakers say one word after a tone or speak conversationally, and whether one person or any person will be able to use the system.

In call centers, speech recognition's biggest use is to replace the touchtone keypad in IVR applications. Using speech recognition opens the door to people who do not have touchtone telephone service, and to older people, who seem to take offense at interacting with their telephone this way. It also eliminates the use of lengthy menus whose only purpose is to convert a word or alphabetic entry into a digit that can be more easily entered into the system.

**SPEED BUMPS** When your host system processes information faster than your network can handle and forces you to slow down.

**SPIFF** A giveaway. An item, usually inexpensive, that is given to call center agents as an incentive. The prize in a call center contest or a reward for achieving a quota.

**SPIKE** 1. A sudden increase in the number of calls coming into a call center or network of call centers. 2. A power fluctuation, typically a sudden increase in voltage.

**SPLIT** An ACD routing division that allows calls arriving on specific trunks or calls of certain transaction types to be answered by specific groups of employees. Also referred to as gate or group.

**SPLIT TEST** A head to head comparison of two or more variations of a marketing campaign. The variants can be different lists (usually samples of different lists), different packages or presentations, or a different offer.

**SPRAY AND PRAY** A habit of help desk clients. When given several numbers to call for assistance (for example, one number for printers, another for software, a third for modems), many people seeking assistance will call and leave messages at all the numbers, hoping to receive assistance faster. The record-keeping problem for your help desk is now you have three cases open, when you really only have one problem to solve.

**SQL** See STRUCTURED QUERY LANGUAGE.

**SS7** See SIGNALLING SYSTEM 7.

**SSP** Service Switching Point. A network-level telephone switch that asks a central database for call handling information.

**STAFFING REQUIREMENTS FORECAST** A calculation of the number of employees required in each period of the day to handle the forecast call volume for that period.

**STAGE & TEST** A term used in the secondary telecom equipment business. The installation (stage) and diagnostic testing of a switch, cabinet, part, or peripheral in a reconfiguration center facility — where a dealer tests the complete system as one entity before shipment.

**STAND ALONE** Any device that can perform independently of something else.

**STANDARDS** Agreed principles of protocol. Standards are set by committees working under various trade and international organizations.

When you're buying a phone system, at minimum it should conform to four standards:

- Emissions compliance according to the FCC Part 15.

- Telephone compliance according to the FCC Part 68.

- Safety standards set by the National Electric Code, OSHA and the Underwriters Laboratories 1459.

- Bellcore compliance (from the Network Equipment Building System publication and their Generic Physical Design Requirements for Telecommunications Products and Equipment publication.

**STARSET** The product name for a Plantronics headset with a contour fit, a voice tube and an ear plug. It's been around for years and continues to be popular in call centers and for receptionists.

**START TIME INTERVAL** A scheduling rule that governs the times at which schedules can start; for example, at 15-minute intervals as opposed to 30-minute intervals.

**STATION** A dumb word for a telephone. Also called an instrument, or a telephone instrument. It's thought that the word comes from the very old days when the telephone industry was regulated by the Interstate Commerce Commission, (the ICC) which also regulated the railroad industry.

**STATION MESSAGE DETAIL RECORDING (SMDR)** See SMDR.

**STATION MESSAGE REGISTERS** Message unit information centrally recorded on a per-station basis for each completed outgoing call.

**STATION SET** The official telecom name for a telephone that works with a phone system or a specific telephone attached to a specific telephone system. Sometimes a station set is proprietary. That is, it will only work with that phone system and won't work at your house or with another telephone system. Where you see "station set" read "telephone."

**STATISTICAL HISTORY** Counts of calls and the elapsed time of call states by trunk group, agent group, trunk port, instrument sign-on, route, DID/DNIS digits, and wrap-up code; a subset of historical data.

**STATISTICS** The science of collecting, tabulating, analyzing and interpreting numerical data. That can be data collected from market research studies, from the reports generated by an ACD or automated dialer, or from sales figures.

**STATISTICS PORT** In network management systems, interface for reporting events and status.

**STATUS DISPLAY** A full-color, graphical image that provides supervisors and call center managers with an instant big picture of real-time conditions. See STATUS MONITORING.

**STATUS MONITORING** The feature of an ACD that shows a supervisor (or call center manager) details of what's happening on a CRT screen. ACDs can monitor trunk usage, agent availability, and peak call volumes, and can measure those statistics against preset tables. When too many calls come in to too few agents, the ACD can be preset to shunt some off to another agent group, or it can be up to the discretion of the supervisor, who sees it as it's happening on the screen.

**STOVEPIPING** In a call center, agents typically need access to many databases. In the past they've used dumb terminals. They log into one computer, get into one database, go further into it. When they need information out of another database, they've typically had to climb out of the previous database, the previous computer, log into another and climb down into it. This is called stovepiping, because it follows the contours of a stovepipe. These days, agents have intelligent computers as terminals. They can access several databases at once, by simply having different windows open on their screen or having a front end program that populates a screen with information from several databases, most likely using a GUI interface.

**STP** Signal Transfer Point. A switch that directs questions about call handling to the appropriate database for 800 and other special calls.

**STRUCTURED QUERY LANGUAGE** SQL. A relational database language that consists of a set of facilities for defining, manipulating and controlling data.

**STRUCTURED WIRING** As data flows sped up in recent years, the idea of wiring up a building with plain old analog voice telephone wire became increasingly not viable. The idea then came along of defining wiring standards and schemes so that a user could feel comfortable about choosing a complete solution for wiring phones, workstations, PCs and other communicating devices throughout the building, the campus, the network, the company. Consistency of design, layout and logic are the keys to structured wiring systems. A structured cabling system will improve performance in five ways, according to Anixter, a leading supplier of structured wiring systems:

1. It eases network segmentation, the job of dividing the network into pieces to isolate and minimize traffic, and thus congestion.

2. It ensures that proper physical requirements, such as distance, capacitance, and attenuation are met.

3. It means adds, moves, and changes are easy to make without expensive and cumbersome rewiring.

4. It radically eases problem detection and isolation.

5. It allows for intelligent, easy and computerized tracking and documentation.

**SUBTARGETING** Marketing from segments of a large database, splitting out names on the basis of desirable factors, like age, income, region, gender or purchasing habits.

**SUPERVISED TRANSFER** A call transfer made by an automatic device such as voice response unit which attempts to determine the result of the transfer — answered, busy, ring no answer — by analyzing call progress tones on the time.

**SUPERVISOR** The person responsible for regulating call flow in and out of a group of agents. On a network, the person who's job it is to keep the system up and running.

**SUPPORT DESK** Another term for help desk. A support desk is a place where incoming calls relating to customer problems are fielded and answered. It is also, more generally, the place where technical experts search for solutions, a key difference between this and other kinds of call centers. A support or help desk may have much non-phone work going on.

**SUPPRESSION FILE** A list of names to be subtracted from a list for a particular campaign. It might be your in-house do-not-call list, the DMA's TPS list or a combination of those and others.

**SURGE PROTECTOR** One of the most important pieces of equipment you can own. It's a device that sits between the computer or phone system and the power outlet.

It protects the expensive equipment from spikes and surges (which are very common and which might reduce your $5,000 computer into $50 worth of scrap metal).

**SUSPECT** A person you believe (for whatever reason) may be interested in your product or service. They may have asked for information about it or they may have purchased a related product or service in the past.

**SVL** Service level.

**SWITCHED ACCESS** A telecommunication service that can be reached by dialing a telephone or dialing through the regular telephone network. For example, a long distance service can be reached through switched access, or it may require a dedicated line or special service such as T-1 for access. Can also refer to data circuits.

**SWITCHING SYSTEMS** General term for the equipment that completes calls in a call center, especially the ACD, its switches, and the trunk lines that connect them to the outside world.

**SWITCHLESS RESELLERS** A switchless reseller buys long distance service in bulk from a long distance company, such as AT&T, and resells that service to smaller users. It typically gets its monthly bill on magnetic tape, then rebills the bulk service to its customers. A switchless reseller owns no communications facilities — switches or transmission. It has two "assets" — a computer program to rebill the tape and some sales skills to sell its services to end users. The profit it makes comes from the difference between what it pays the long distance company and what it is able to sell its services at. Switchless resellers are also called rebillers or aggregators. It's not an easy business to be in, since you are selling a long distance company's services to compete against itself.

**SYNTHESIZED SPEECH** A speech message created by a computer from basic sound parts. It sounds funny, like a drunken Swede is reading the message. Messages assembled from whole words or phrases recorded by a human being usually provide higher quality (when you call an automated system for your bank system or when you call directory assistance, this last method is what they use.)

Synthesized speech is best used when each message will be so unique that prerecording words or phrases will not work.

**SYSTEMS INTEGRATOR** A consulting organization (or consulting arm of a product vendor) whose role is to coordinate the installation and implementation of complex call center and computer telephony projects. Integrators are typically called in when a project is large and includes custom elements, like connections to legacy host systems and databases, or phone switches for which connections still need to be written. They may also be needed when the call center applications themselves are complex and custom, especially in the financial services sector. Systems integrators often have a role in product recommendation and selection, as well.

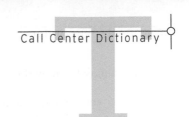

**T.120** A series of standards that let you share an on-line document in real-time with several users. Because these standards were created by the ITU-T (International Telecommunications Union), a United Nations-based organization, they can be used by any vendor. These are the standards that let you look at a computer screen along with several other people at other locations and draw all over it. The standards let you send files and work on multisite "whiteboards." When you hear about multimedia call centers where agents and customers can fill out forms together or edit a word processing file in real-time together, the T.120 standards are what's behind it all.

**T-1** Also spelled T1. A digital transmission link with a capacity of 1.544 Mbps (1,544,000 bits per second). T-1 uses two pairs of normal twisted wires, the same as you'd find in your house. It normally can handle 24 voice conversations, each one digitized at 64 Kbps. But, with more advanced digital voice encoding techniques, it can handle more voice channels. T-1 is a standard for digital transmission in the United States, Canada, Hong Kong and Japan.

T-1 lines are used for connecting networks across remote distances. Bridges and routers are used to connect LANs over T-1 networks. There are faster services available. T-1 links can often be connected directly to new PBXs and many new forms of short haul transmission, such as short haul microwave systems. It is not compatible with T-1 outside the United States and Canada. In Europe T-1 is called E-1 or E1.

Outside of the United States and Canada, the "T-1" line bit rate is usually 2,048,000 bits per second. France and West Germany impose slight variations that make their formats unique. According to Bill Flanagan, the differences are not so great that a multiplexer cannot convert between them.

At the higher rate of 2,048,000, 32 time slots are defined at the CEPT interface, but two are used for signaling and other housekeeping chores. Typically 30 channels are left for user information — voice, video, data, etc. CEPT is the Conference of European Postal and Telecommunications administrations, a standards-setting body whose membership includes European Post, Telephone, and Telegraphy Authorities (PTTs).

For a full explanation of T-1 see Bill Flanagan's book The Guide to T-1 Networking.

**T-1 FRAMING** This is a rather technical explanation of a critical function of call center planning — the passing of call information over high speed lines. Bear with it, it comes in handy to know this:

Digitization and coding of analog voice signals requires 8,000 samples per second (two times the highest voice frequency of 4,000 Hz) and its coding in 8-bit words yields the fundamental T-1 building block of 64 Kbps for voice. This is termed a Level 0 Signal and is represented by DS-0 (Digital Signal at Level 0). Combining 24 such voice channels into a serial bit stream using Time Division Multiplexing (TDM) is performed on a frame-by-frame basis. A frame is a sample of all 24 channels (24 x 8 = 192) plus a synchronization bit called a framing bit, which yields a block of 193 bits. Frames are transmitted at a rate of 8,000 per second (corresponding to the required sampling rate), thus creating a 1.544 Mbps (8,000 x 193 = 1,544 Mbps) transmission rate, the standard North American T-1 rate. This rate is termed DS-1.

**T-2** A link that passes information at the rate of 6.312 million bits per second. Capable of handling at least 96 voice conversations depending on the encoding scheme chosen. T-2 is four times the capacity of T-1. T-2 is further up the North American digital carrier hierarchy. See T-1.

**T-3** 28 T-1 lines or a link that passes information at the rate of 44.736 million bits per second. Commonly referred to as 45 megabits per second. Capable of handling 672 voice conversations. T3 runs on fiber optic and is typically called FT3. T-3 is further up the North American digital carrier hierarchy. See T-1.

**T-4** A link that passes information at the rate of 274.176 million bits per second. Capable of handling 4032 voice conversations. T-4 has 168 times the capacity of T-1. T-4 can run on coaxial cable, waveguide, millimeter radio or fiber optic. T-4 is further up the North American digital carrier hierarchy. See also T-1.

**TABLES** A collection of data in which each item is arranged in relation to the other items. Many telephony functions use "look up tables" to determine the routing of calls. These tables solve the problem, "If the call is going to this exchange in this area code, then use this trunk and this routing pattern."

**TALK TIME** The length of time agents spend placing or answering calls (as opposed to the time between calls that they spend updating records, sending out literature, or going to the bathroom.) In outbound call environments where agents dial out manually, typical talk time is close to 20 minutes per hour. When you use automation to increase the number of calls that actually reach their target, talk time can more than double. Interestingly, reports from the field indicate that when you use automation to increase agent talk time, agents actually enjoy their job more, burn out less often, and stay employed longer. The upshot: even more productivity than you would think at first glance, due to lower training costs and higher morale.

**TALK-OFF** When your voice has enough of the specific frequency used by the telephone network or a voice processing system to signal a call is disconnected that the system actually disconnects you.

**TAPI** TAPI stands for Telephony Application Programming Interface. It's Microsoft's protocol for linking Windows applications to telecommunications devices. If you are not a systems developer you may never actually use TAPI, but since it seems impossible to get away from Windows in your call center these days, you should know the term.

**TARGETED SUBSETS** The groups of potential customers that have been segmented out of a larger list or database. You can tailor marketing campaigns and appeal to specific characteristics of the members of a group.

**TARIFF** A document filled by AT&T (the only regulated long distance company) with the Federal Communications Commission or by a local telephone carrier with its state public utilities commission. MCI and Sprint also voluntarily file tariffs with the FCC.

If accepted, a tariff regulates the prices or services that a carrier can offer.

**TARIFF 12** While tariffs are supposed to assure that everyone serviced by a carrier gets the same price, Tariff 12 lets AT&T set special prices for large long distance contracts. This is so they can fairly compete on these contracts with MCI and Sprint, who are not held to their tariffed rates by the FCC. You just can't make stuff like this up.

**TARIFF 15** A user-specific long distance tariff of AT&T. Tariff 15 gives AT&T the ability to price its long distance services for one company practically any way it feels. Tariff 15 is single-customer discounting. Some of AT&T competitors claim the tariff is "illegal."

**TCAP** Transactional Capabilities Application Part. Provides the signaling function for network databases. TCAP is an ISDN application protocol (one of three).

Some examples of services made possible by this protocol include enhanced toll-free service, automated credit card calling and virtual private networking.

The TCAP protocol lets these services access remote databases called service control points (SCPs) to process part of the call. The SCP supplies the translation and routing data needed to deliver advanced network services.

**TCP/IP** It stands for "Transmission Control Protocol/Internet Program" and it's a computer networking term. TCP/IP is a protocol — commonly used over Ethernet and X.25 networks — that lets different types of computers on different types of networks talk to each other.

This protocol has four layers: a network interface layer, an Internet layer (yup, that Internet), a transport layer and an application layer. These layers correspond to some of the seven layers in the famed, but not often used, Open Systems Interconnection (OSI) standards.

TCP/IP was developed by the military, but its success in linking systems from different vendors makes it popular with commercial users too.

**TDD** Telecommunications Device for the Deaf. Long before personal computers existed it occurred to someone that deaf people could communicate by telephone if only they could use print instead of speech. The existing print-by-telephone technology was teletypewriters (TTY). These were adapted to create TDDs. TDDs can resemble a typewriter or a small computer. They have keyboards and a printer or screen.

Now that personal computers are common, special modems are available to translate computer messages into TDD messages and visa versa.

**TECHNICAL SUPPORT** The group at a customer service center that actively seeks out solutions to customer problems. This is a varied group of people. Technical support, from the customer's point of view, includes the people who sit on the phone, answer questions, and look up solutions in databases and manuals. But it also includes the people who work behind the scenes to solve problems, and the technicians who are dispatched for onsite service.

**TECHNO-GEEK** A person who loves technology for technology's sake and needs to be on the cutting edge, even if it's not practical or cheap. Usually knows all the tiny details of technology and looks down on those who don't.

**TELCO** The local telephone company. Can refer to either a Bell company or one of the over one-thousand other companies that own or run a local telephone company.

**TELECOMMUNICATIONS DEVICE FOR THE DEAF** See TDD.

**TELECOMMUNICATIONS INFRASTRUCTURE** The services and physical telecom technology that's available in any given area. This is of particular concern when choosing where to locate a call center. You need to assess the access to long distance lines, ANI, and T-1 lines, for example. Also important is proximity to a network's point of presence from the local phone company.

**TELECOMMUTING** The practice of working from home or a location remote from your office. Call center agents are particularly well-suited to the practice of telecommuting and new technology means that agents can work from home even more efficiently than they could ten years ago. Unfortunately, many call centers and other businesses tried telecommuting ten years ago and couldn't get it to work. Because of a lingering bad taste associated with "telecommuting," vendors of the new technology (which often uses the Web) now shun the term "telecommuting" and use instead "work at home." They mean the same thing.

**TELECOMPETITIVENESS** A measurement of how well one potential call center location stacks up against another. Different people measure telecompetitiveness differently. Some of the things that might make up a telecompetitiveness index are:

telecom infrastructure; government regulations; tax rates; and the labor pool (quality and quantity).

**TELEMANAGEMENT** Software and systems that help you keep track of and control the telecommunications in your business. Traditionally, Telemanagement comprises call accounting (tracking the flow of calls in and out) and facilities management (tracking the "physical plant" of equipment and cable locations, terminations, and inventory).

Telemanagement includes every function the corporate telecommunications manager does today — from ordering new circuits, to managing the corporate inventory of phones, lines and other equipment, to choosing the right number of agents to staff the phones at the right time. Increasingly, you now see personal computers sitting on top of telephone systems, collecting and processing information. That's part of telemanagement, too.

**TELEMARKETING** The art (or science) of selling products and service over the phone. Although it comes in two flavors (inbound and outbound), outbound is by far the first thing that people mean when they use the term.

Outbound telemarketing is further broken down into two components: business-to-business and business-to-consumer. The two are distinguished by the time of day you are likely to make the sales call (business hours vs. early evening), and to some extent the kind of technology used to automate the telemarketing project. You are much less likely to use a predictive dialer, for example, on a business-to-business campaign than you are on a consumer campaign. The reason is that you are much more likely to reach voice mail or an automated attendant when calling a business, nullifying the benefits of predictive's look-ahead dialing.

Inbound telemarketing is largely run through toll-free numbers. The main examples of inbound telemarketing are catalog retailing, travel reservations, and financial service transactions.

**TELEPHONE** 1. An invention of the devil. 2. The biggest time waster of all time, as in: "What did you do all day?" "Nothing. Just spent the day on the phone."

**TELEPHONE CONSUMER PROTECTION ACT** A Federal law, enforced by a set of FCC regulations, effective in December 1992, that requires companies to keep a list of consumers who have requested not to receive phone solicitations from their company. Other provisions of this act say companies can't call consumers at home between 9 PM and 8 AM; companies must obtain consumer's consent to share his or her phone number with other marketers; and that automated message players can not call emergency lines, health care facilities and numbers where the other party pays for the call, like cellular phones.

**TELEPHONE MANAGEMENT SYSTEM** The term originally meant a system for controlling telephone costs by: 1. Automatically selecting lower-cost long distance routes for placed calls; 2. Automatically restricting certain people's abilities to make some or all long distance calls; and 3. Automatically keeping track of telephone usage by extension, time of day, number called, trunk used and sometimes by person calling and client or account to be billed for call.

These days this term is collapsed into the umbrella of TELEMANAGEMENT. That term covers both the equipment used in the process and the process itself.

**TELEPHONE NUMBER APPENDING** Adding telephone numbers to mailing lists.

**TELEPHONE PREFERENCE SERVICE** A service provided by the Direct Marketing Association (DMA). It is a national list of people who request that their names be taken off telemarketing lists. It is a good idea to have your marketing lists checked against the TPS before embarking on a campaign.

**TELEPHONE SERVICE REPRESENTATIVE TSR.** Another word for agent — the person who or makes telephone calls in a call center. See AGENT.

**TELEPHONE SET EMULATION** The process by which you put all the functions of a physical telephone onto a printed circuit card, and add that card into a PC. Why would you want to do that? Because it saves time adds possibilities.

The PC does everything a human using the phone could do, only the PC will do it more efficiently. The human will find it easier to use all his phone's features because the PC's screen is bigger and the PC's keyboard easier to use than the phone's keypad.

And once inside the PC, the (once-proprietary) phone can be attached to voice and call processing cards, like voice recognition, voice mail, touchtone generation, and bingo, phone systems acquire all the benefits of integrated voice and call processing.

**TELEPHONE USE RESTRICTION** A direct mail list that can't be used by telemarketers — as specified by the list owner.

**TELEPHONE-BASED SPEECH RECOGNITION** Sometimes called telephony-based. Speech recognition that occurs over the telephone. It is important for call center managers to know that telephone-based speech recognition makes up just 30% of the speech recognition market (according to Probe Research). The telephone makes speech recognition more difficult, because of line noise, background noise, and differences in handsets and headsets. When you hear of some marvelous speech recognition application or feature, make sure you know whether it works over the telephone or not.

**TELEPHONY** The science of transmitting voice, data, video or image signals over a distance greater than what you can transmit by shouting.

For the first hundred years of the telephone industry's existence, the word telephony described the business the nation's phone companies were in. It was a generic term. In the early 1980s, the term lost fashion and many phone companies decided they were no longer in telephony, but in telecommunications — a more pompous sounding term that was meant to encompass more than just voice.

In the early 1990s, as computer companies started entering the telecommunications industry, the word telephony was resurrected. And in a white paper on multimedia from Sun Microsystems, the company said that telephony refers to the integration of the telephone into the workstation.

For instance, making or forwarding a call will be as easy as pointing to an address book entry. Caller identification (if available from the telephone company) could be used to automatically start an application or bring up a database file. Voice mail and incoming faxes can be integrated with e-mail (electronic mail). Users can have all the features of today's telephones accessible through their workstations, plus the added benefits provided by integrating the telephone with other desktop functions. See also COMPUTER TELEPHONY.

**TELEPHONY SERVER** A telephony server is a computer whose major function is to control phone calls (in all their permutations of fax, voice and data).

The traditional function of a telephony server is to move call control commands from client workstations on a LAN to an attached PBX or ACD. (This is what it does under the paradigm called "Telephony Services.") Examples of telephony servers can include voice response systems and fax-on-demand systems.

**TELEPHONY SERVER** A computer whose main function is to control or manipulate the various calls (voice, fax, e-mail, IVR or other data) flowing into and out of a computer telephony system, usually as part of a call center. The traditional role of a telephony server is to move call control commands from client workstations on a LAN to an attached ACD or PBX. More important, the telephony server is now seen as a way to roll multiple telephony devices — IVR, fax-on-demand, voice processing, auto attendant — each of which used to sit in its own box, into one machine.

**TELEPHONY SERVICES** Telephony Services's real name is Telephony Services for NetWare. NetWare is the name of the best-selling local area network software supplied and sold by Novell. Telephony Services for NetWare basically consists of an addition to the NetWare operating system, called Telephony Server NLM. That addition handles communications between a NetWare file server and an attached telephone switch, e.g. a PBX or ACD.

Telephony Services for NetWare is basically the software in the NetWare file server which takes care of interpreting your PC commands into commands your switch can understand and respond to.

Telephony Services requires a link to your switch. Each telecom switch manufacturer has been implementing that link in a different way. That's fine, because Telephony Services for NetWare insulates the user and the developer. This means that computer telephony applications written for Telephony Services for NetWare will work on any switch conforming to the Telephony Services standard. As of this writing, virtually every switch manufactured in North America and many made in Japan and Europe is conforming to Novell's Telephony Services.

According to a White Paper issued by Novell in March 1994 and called NetWare Telephony Services, "The three main components of Telephony Services are call control, voice processing, and speech synthesis."

Client/server computer telephony (which is what all this is called) delivers these benefits to call centers:

1. Synchronized data screen and phone call (known popularly as "screen pop" or "screen synch"). Your phone rings. The call comes with the calling number attached (via Caller ID or ANI). Your ACD passes that number to your server, which does a quick database lookup to see if it can find a name and database entry.

If it does, it passes the call and the database entry simultaneously to the agent. It saves the customer asking a lot of questions. Makes customers happier.

2. Database transactions. Customer lookups of bank account balances, airline reservations, catalog requests, movie times, etc. Doing business over the phone is exploding. Today, the caller inputs his request by touchtone or by recognized speech. The system responds with speech and/or fax.

Today's systems are limited in size and flexibility. Add the power of a LAN, and suddenly you've suddenly got a computer telephony system that knows no growth constraints. You could also get the system to front-end an operator or an agent. Once the caller has punched in all his information, then the call and the screen can be simultaneously passed to the agent.

3. Telephony work groups. Sales groups. Collections groups. Help desks. R&D. We work in groups. But traditional telephony doesn't. Groups have special needs.

For example, the company's help desk needs a front end voice response system that asks for the customer's serial number, some indication of the problem and tries to solve the problem by instantly sending a fax or encouraging the caller to punch his way to one of many canned solutions.

When all else fails, the caller can be transferred to a live human, expert at diagnosing and solving his pressing problem.

Or a development group might need e-mails and faxes of meeting agendas sent, meeting reminder notices phoned and scheduled video conferences set up. All automatically. The accounts receivable department needs a predictive dialer to dial all overdue customers. The telephone sales department also needs a predictive dialer, but with different programming.

4. Management of phone networks. Today, phone networks are very difficult to manage. Often the PBX is managed separately from the voice mail, which is managed separately from the call accounting, etc. It's a rare day in any corporate life when the whole system is up to date, with extensions, bills and voice mail mailboxes reflecting the reality of what's actually happening.

Integrate LAN management tools with telecommunications management, and potentially all you need is to make one entry (for a new employee, a change, etc.) and the whole system — telecom and computing — could update itself automatically, including even issue change orders to the MIS and telecom departments and vendors.

5. Switch elimination. The ultimate potential advantage of LAN-based telephony is to eliminate the connection to the switch (PBX or ACD) by simply populating the LAN server (now called a telephony server) with specialized computer telephony cards and run the company's or department's phones off the telephony server directly.

**TELESCRIPTS** This term, now a somewhat generic word applied to any automated telemarketing scripts, also refers to: an old telemarketing software program from a company called Digisoft, and a feature of Rockwell switches for configuring the many options of routing a call in progress, a call in queue, a call in the system, and so forth.

**TELETRAFFIC OPTIMIZER PROGRAM** A forecasting system that derives data by processing actual calls instead of using an analytical model based on estimates or summaries.

**TELEWORKER** A person who works from his home or some place distant from his company's office. A teleworker may send his completed work in and pick his new work up via a modem in his PC. A teleworker may also be on the phone at home answering calls on behalf of his company and entering the results of those calls (i.e. reservations on airlines, orders for catalogs) in on a PC connected by phone lines to his company. He may use one phone line, like an ISDN BRI line or he may simply use two analog phone lines — one for talking on and one for PC's data. Or he may simply use one analog phone line and use a protocol such as VoiceView. Call centers are increasingly relying on Teleworkers (also known as telecommuters or "agents-at-home") as a way to keep from being overloaded during unforeseen peaks in call volume.

**TEMPORARY SIGNALING CONNECTIONS** On August 14, 1995, AT&T announced Temporary Signaling Connections, which it billed as the first service that lets banks, retail outlets and other data-intensive businesses link their Software Defined Network locations together on demand using virtual connections created in AT&T's national signaling networks. Businesses can use Temporary Signaling Connections to verify credit card transactions, update inventory databases, exchange data with automatic cash machines. The service uses a portion of the D channel capacity of an ISDN PRI channel and passes information to other ISDN PRI locations.

**TEXT-BASED RETRIEVAL** A method of finding answers to customer service problems. It is the simplest of automated methods. All problems and solutions that the help desk has recorded are stored in one or more large text files. Solutions are found by querying that descriptive text. A feature or function of help desk or technical support software systems.

**TEXT-TO-SPEECH** A system that converts information from a computer database into spoken information. Usually refers to a system that uses SYNTHESIZED SPEECH.

**THIN CLIENT** A stripped down PC that does only the basics. The "dumb terminal" of the world of computer client/server networks.

**THIRD PARTY CALL CONTROL** A call comes into your desktop phone. You can transfer that call. When the phone call has left your desk, you can no longer control it. That is called First Party Call Control. If you were still able to control the call (and let's say, switch it elsewhere) that would be called Third Party Call Control. If you control the switch — the PBX or the ACD — you will typically have Third Party Call Control. If you just control the desktop, you'll typically have only First Party Call Control.

**THRESHOLD** Also referred to as an "alert threshold." Literally, the point at which something happens. In call centers, targets are often set for things like performance or service level. A center might aim for answering 80% of calls within 20 seconds, for example (a common number). And systems are put in place to track how long those calls are actually waiting. But it's not good enough to know after the fact that you did, or didn't make your target. It's important to know as it's happening, so you can take some action.

Lots of systems allow the user to set thresholds - to set points at which the system delivers a warning to an agent or a supervisor. If a call is in the queue for more than ten minutes, for example, it might sent a note of that fact to the supervisor terminal. A status display might turn yellow, or red, as thresholds are passed during a center's activity.

You can set thresholds for your workforce management system, to tell you when individual and group performance is not where it should be. Or you can set them in the call routing system - to tell you, for example, when calls should be diverted from one call center to another.

**THROUGHPUT** The rate of total transfer of information over communications lines. It differs from the actual speed. For example, say you are transferring a file at 9,600 bits per second over a modem. The throughput of that transfer is 9,600 bps. But if you compress the file first, reducing its length by half (and then decompress it on the other side), the actual throughput is 19,200 bps.

**TIME ZONE REPORT** A report that shows the number of phone numbers left to dial by time zone.

**TIME ZONE SEQUENCING** Putting a telephone calling list in time zone order. Important because telemarketing is national; one way to maximize resources in your center is to set up your staff and your campaigns so that you're processing a list in rolling order across time zones.

**TIME-OF-DAY ROUTING** A way to distribute calls between agent groups (or even between unique call centers). Time-of-day routing refers to several aspects of call distribution:

1. Sending calls after a certain hour from one group to another; to cover for lunch breaks or scheduled meetings, for example.

2. Setting a threshold hour when calls stop going to one group or center, and instead go in another direction. When the office closes, for example, and you want to send calls to a center in a region where the time zone differential keeps that center open later. Sending calls from the New York office to the Los Angeles office after 5 pm New York time, while it's still 2 pm on the West coast.

Time-of-day routing also applies when you send those post-5 PM calls that arrive in New York to a voice processing system instead of to an actual agent group.

3. Using the network intelligence supplied by the toll-free carriers to change the way incoming calls are supplied to your center. Bringing in calls on MCI for one period of time, then switching to AT&T for another, predetermined time frame.

**TIMECLOCK** One thing you'll find in reporting packages from ACD vendors is a "timeclock" function. This function uses ACD statistics to compare staff schedules to the actual times agents worked.

These packages print out an exception report telling you who signed on five minutes or more past their scheduled start time — even if he or she makes up the time by working five minutes later. You find out exactly when your agents worked, not just that they all put in eight hours.

If even your best schedules don't cover your calls, the timeclock function could tell you if scheduling deviations are the cause.

**TIMED REMINDERS** At 20-second intervals, timed reminders will alert an agent that a call is still waiting, a called line has not yet been answered or a call is still on hold. Timed reminders can be made longer or shorter. They can alert agents to all sorts of events and non-events.

**TOKEN RING** A communications protocol that allows the workstations of a network to send packets of data (the "token") around a logical ring of stations. Any station that wants to transmit must have possession of the token in order to do so.

**TOLL FRAUD** Theft of long distance service. According to John Haugh of Telecommunications Advisors in Portland, OR, there are three distinct varieties of toll fraud:

"First Party" Toll Fraud, which is helped along by a member of the management or staff of a user. An example would be the telecommunications manager at the Human Resources department of New York City (an "insider") who sold his agency's internal code to the thieves, who in turn ran up unauthorized long distance charges exceeding $500,000.

"Second Party" Toll Fraud, which is facilitated by a staff member or subcontractor of a long distance carrier IXCs, vendor or local exchange telephone company selling the information to the actual thieves, or their "middlemen." An example would be a "back office clerk" working for one of these concerns who sells the codes to others.

"Third Party" Toll Fraud is facilitated by unrelated "strangers" who, though various artifices, either "hack" into a user's equipment and learn the codes and procedures, or obtain the needed information through some other source, to commit Toll Fraud.

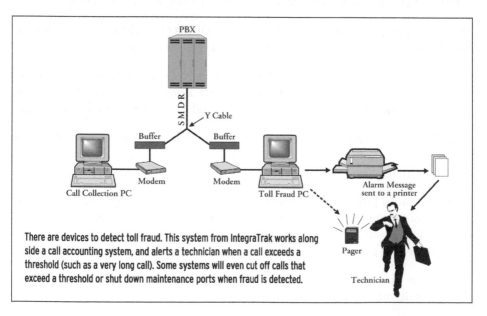

There are devices to detect toll fraud. This system from IntegraTrak works along side a call accounting system, and alerts a technician when a call exceeds a threshold (such as a very long call). Some systems will even cut off calls that exceed a threshold or shut down maintenance ports when fraud is detected.

**TOLL FREE SERVICE** A bit of a misnomer, since the call is not free, but is payed for by the party that receives the call. That's why it is also called "called party paid service." We put the main definition here because that's too much of a tongue twister. This service was called "800 service" until the 888 toll-free exchange was introduced. Old-timers may also know this service as "inbound WATS."

Toll free service works like this: You dial 1-800 or 1-888 (or 1-877) and seven digits from somewhere in North America. Your local central office sees the "1" and recognizes the call as long distance. It ships that call to a bigger central office. At that central office a machine recognizes the 800/888/877 "area code" and examines the next six digits, and performs a database lookup which tells it which long distance carrier will handle the call. Those six digits also tell the carrier where to send the call.

Because of the sophistication of the database, there are many routing variations available for any toll-free number. Routing can change by day or time of day, a traffic-volume threshold, location the call originated from or by percentage of calls.

Here is how a toll-free database lookup works, according to Bellcore. The brackets show changes we made to update their description to the current state of affairs.

"The telecommunications network architecture that supports 800 Data Base Access Service is considered "intelligent" because data bases within the network supplement the call processing function performed by network switches. The Service uses a Common Channel Signaling (CCS) network and a collection of computers that accept message queries and provide responses. When a caller dials an 800 [or 888] number, a Service Switching Point (SSP) recognizes from the digits "8-0-0" [or "8-8-8"] that the call requires special treatment and processes that call according to routing instructions it receives from a centralized database. This database, called the Service Control Point (SCP), can store millions of customer records.

"Information about how [a toll-free] call should be handled is entered into the SCP through the off-line Bellcore support system called the Service Management System (SMS). SMS is a national computer system which administers assignment of toll-free numbers to toll-free service providers."

As of this writing, the database is located in Kansas City, maintained by the Database Service Management (DSMI), a wholly-owned subsidiary of BellCore, with day-to-day administration performed by a division of Lockheed. DSMI is trying to get the physical database relocated to New Jersey.

Toll free service is regulated by the Federal Communication Commission (FCC), which relies on recommendations from a large number of industry forums. ATIS (Alliance for Telecommunications Industry Solutions) is another important player. Several key industry committees or forums, such as the Industry Numbering Committee (INC), the Ordering and Billing Forum/Service Management System/800 Number Administration Committee (OBF/SNAC) forum and the Network Operations Forum (NOF) operate under the ATIS umbrella. Another important committee is the Carrier Liaison Committee.

Here are a few examples of the services toll-free service providers have created using the variables allowed with the database:

- TIME OF DAY ROUTING: Allows you to route incoming calls to alternate, pre-determined locations at specified days of the week and times of the day.

- PERCENTAGE ALLOCATION ROUTING: Allows you to route pre-selected percentages of calls from each Originating Routing Group (ORG) to two or more answering locations. Allocation percentages can be defined for each ORG (typically an area code), for each day type and for each time slot.

- SINGLE NUMBER: The same toll-free number is used for intrastate and interstate calling.

- CALL BLOCKAGE: You can block calling areas by state or area code. The caller from a blocked area hears the message: "Your toll-free call cannot be completed as dialed. Please check the number and dial again or call 1-800-XXX-XXXX for assistance." (You may want to block callers from areas which didn't see your special commercial, for example.)

- POINT OF CALL ROUTING: Allows a customer to route calls made to a single toll-free number to different terminating locations based on the call's point of origin (state or area code.) You establish Originating Routing Groups (ORGs) and designate a specific answering location for each ORG's call.

- CALL ATTEMPT PROFILE: A special service that allows subscribers to purchase a record of the number of attempts that are made to an toll-free number. The attempts are captured at the Network Control Point, and from this data a report is produced for the subscriber.

- ALTERNATE ROUTING: Allows a customer to create alternate routing plans that can be activated by the toll-free carrier upon command in the event of an emergency. Several alternate plans can be set up using any features previously subscribed to in the main toll-free number routing plan. Each alternate plan must specify termination in a location previously set up during the order entry process.

- DIALED NUMBER IDENTIFICATION SERVICE: DNIS. Allows a customer to terminate two or more toll-free numbers to a single service group and to receive pulsed digits to identify the specific toll-free number called. DNIS is only available on dedicated access lines with four-wire E & M type signaling or a digital interface. The customer's equipment must be configured to process the DNIS digits.

- ANI: The carrier will deliver to you the incoming toll-free call plus the phone number of the calling party. See also ANI, COMMON CHANNEL INTEROFFICE SIGNALING and ISDN.

- COMMAND ROUTING: Allows the customer to route calls differently on command at any time his business requires it.

- FOLLOW ME 800: Allows the customer to change his call routing whenever he wants to.

**TOLL RESTRICTION** A feature of a phone system that curbs a telephone user's ability to make particular calls. Toll restriction capability on modern PBXs and key telephone systems has been increasing in sophistication. Some PBXs now allow selective restriction based on specific extensions, users or geography. You can cut out specific area codes, for example, like 900. Or you can let a sales rep make calls to only one region of the country.

**TONE DIALING** To dial a telephone using a series of dual-tone, multi-frequency sounds (DTMF) to signal the telephone network, phone system or other device (voice mail system, computer). The keypad used for tone dialing has a rectangle of 12 push buttons.

**TONER PHONER** An office supply company that uses shady telemarketing tactics. Often the supplier implies (or just lies about) a prior relationship with the company. The supplier sells low-quality supplies at greatly inflated prices, then refuses returns or disappears before a complaint can be made. The name comes from the fact that these companies originally specialized in photocopy toner, then moved on to fax paper. Something to keep in mind if you plan to sell office supplies by telephone.

**TOP DOWN** This is a method of distributing incoming calls to a bunch of people. It always starts at the top of a list of agents and proceeds down the list looking for an available agent. See also ROUND ROBIN and LONGEST AVAILABLE.

**TOPOLOGY** A term used in STRUCTURED WIRING (the coordination of wiring plans within a call center). The way the cable is physically laid out or configured. Examples include star, ring, daisy chain and backbone.

**TOTAL TRANSACTION CALL PROCESSING** A Rockwell term. It involves managing the success of a call center, not merely supplying the ACD (Automatic Call Distributor). It could include software development, CTI integration, network management, consulting services, IVR and voice processing systems.

**TOUCHTONE** The name AT&T originally gave to their tone dialing keypad and telephone with one of those keypads. They don't seem to enforce the trademark. Also called "tone dial" and "DTMF."

**TPS** See TELEPHONE PREFERENCE SERVICE.

**TRACE AGENT** This is a command used in the Teknekron Infoswitch product line to report all the events and transactions an agent has been involved in over a defined period of time.

**TRACKING** A software feature that models actual events and activities in your call center to aid you in short-term planning and evaluation of employee and call center performance. The tracking functions include employee information scheduling assignment, daily activity, and intra-day performance.

**TRAFFIC** Bellcore's definition: A flow of attempts, calls, and messages. Another way to look at it: The amount of activity during a given period of time over a circuit, line or group of lines, or the number of messages handled by a communications switch. Imagine telephone calls are cars and telephone lines (circuits, trunks) are roads and highways. You can have light traffic or heavy traffic (but the term "bumper to bumper" is not used).

There are many measures of "traffic." Typically it's measured in minutes of voice conversation, or bits of data conversation. Note that Bellcore includes attempts in its definition of traffic. Some do not. The decision is yours. But you should be aware of what you include in your calculations.

**TRAFFIC CAPACITY** The number of CCS (hundred call seconds) of conversation a switching system is designed to handle in one hour.

**TRAFFIC CARRIED** See TRAFFIC OFFERED AND CARRIED.

**TRAFFIC CONCENTRATION** The average ratio of the traffic during the busy hour to the total traffic during the day.

**TRAFFIC DATA TO CUSTOMER** The owner of a call accounting system can poll her PBX daily or hourly and get traffic measurements, including peg counts, usage and overflow data. She can also get summary reports, exception reports and complete traffic register outputs.

**TRAFFIC ENGINEERING** The science of figuring out how many trunks, how much switching equipment, how many phones, and in general how much communications equipment you'll need to handle the telephone traffic you're estimating. Traffic engineering suffers from several problems:

1. You are basing your future needs on past traffic.

2. Most traffic engineering is based on one or more mathematical formulas, all of which approach but never quite match the real world situation of an actual operating phone system. Computer simulation is the best method of predicting one's needs, but it's expensive in both computer and people time.

3. Since there are now several hundred long distance companies in the United States and several thousand differently-priced ways of dialing between major cities (most at different prices), traffic engineering has become very complex.

Traffic Engineering uses probability theory to estimate the number of servers required to meet the needs of an anticipated number of customers. In telephone work, the servers are often trunks, and the customers are telephone calls, assumed to arrive at random. When arriving calls, upon finding all trunks busy, vanish, you get what's called a "blocked call cleared" situation. When a call stays in the system for a given length of time, whether it gets a trunk or not, "blocked calls held" applies. If a call simply waits around until a trunk becomes available and then uses the trunk for a full holding time, the correct term is "blocked calls delayed."

Like any form of predicting the future on the basis of past behavior, traffic engineering has its limitations; however, when used by those who have taken the trouble to learn how it works, its track record is surprisingly good, and vastly better than most forms of simulation.

**TRAFFIC INTENSITY** A measure of the average occupancy of a facility during a period of time, normally a busy hour, measured in traffic units (erlangs). It's defined as the ratio of the time during which a facility is occupied (continuously or cumulatively) to the time this facility is available.

A traffic intensity of one traffic unit (one erlang) means continuous occupancy of a facility during the time period under consideration, regardless of whether or not information is transmitted.

**TRAFFIC LOAD** Total traffic carried by a trunk during a certain time interval.

**TRAFFIC MEASUREMENT** Memory and other software in a telephone system which collects telephone traffic data such as the number of attempted calls, the number of completed calls and the number of calls encountering a busy. The objective of traffic measurement is to enter the results into a traffic engineering calculation and so arrange one's incoming and outgoing trunks to get the best possible service.

**TRAFFIC MONITOR** A PBX feature that provides basic statistics on the amount of traffic handled by the system.

**TRAFFIC OFFERED AND CARRIED** People pick up the phone and try to place their calls. This is "Traffic Offered" to the switch. The calls that get through the switch and onto lines is called "Traffic Carried." The difference between traffic offered and carried is the traffic that was lost or delayed because of congestion. There are two basic ways of measuring traffic — erlangs and CCS (or hundred call seconds).

**TRAFFIC OVERFLOW** The condition that occurs when traffic flow exceeds the capacity of a particular trunk group and flows over to another trunk group.

**TRAFFIC PATH** A path over which individual communications pass in sequence.

**TRAFFIC RECORDER** A device which measures traffic activity on a transmission channel. It does no processing, only observation of conditions.

**TRAFFIC TABLE** A computer database into which a switch enters a count of activity. Certain detected operating errors are also entered in the traffic table.

**TRAFFIC USAGE RECORDER** A device for measuring and recording the amount of telephone traffic carried by a group, or several groups, of switches or trunks.

**TRAINING (VOICE RECOGNITION)** Speaker-dependent voice recognition systems must be "trained" to recognize their master's voice. The user reads a word or phrase into the system several times. The system uses these repetitions to create a template that will determine future recognition.

**TRANSLATIONS** Translations are changes made by the network to dialed telephone numbers to allow the call to progress through the network. Sometimes the translations are made automatically. Take one series of dialed numbers; convert them to another. Sometimes, translations are done with the help of "lookup" tables, also called databases.

**TRANSLATOR** 1. In telephone equipment, it is the device that converts dialed digits into call-routing information.

2. A communications device that receives signals in one form, normally in analog form at a specific frequency, and retransmits them in a different form.

3. A device that converts information from one system into equivalent information in another system.

**TRANSMISSION GAIN** Most electronic telephones require that headsets boost their transmissions to various levels for the best sound quality. A transmission gain adjustment lets you fine-tune your headset for your particular telephone. This term means something completely different when referring to telecommunications media, such as microwaves.

**TRANSPARENT** Not apparent to the user or caller. Seamless. Requiring no action, though or even awareness on the part of the user or caller. Used to describe a technology, especially a network or interface between two or more technologies. For example, when you call an IVR system for a bank balance and you only have to identify yourself through your PIN number, the ANI-based database lookup that retrieved your banking record was transparent to you, as the caller.

**TRUNK** A communication line between two switching systems. The term switching systems typically includes equipment in a central office (the telephone company) and your customer-based equipment (PBX or ACD).

**TRUNK ACCESS NUMBER** The number of the trunk over which a call is to be routed.

**TRUNK GROUP** A group of essentially like trunks that go between the same two geographical points. They have similar electrical characteristics. A trunk group performs the same function as a single trunk, except that on a trunk group you can carry multiple conversations.

**TRUNK GROUP ALTERNATE ROUTE** The alternate route for a high-usage trunk group. A trunk group alternate route consists of all the trunk groups in tandem that lead to the distant terminal of the high-usage trunk group.

**TRUNK HOLDING TIME** The length of time a caller is connected with a voice processing system. Defined from the time when the system goes off-hook to the time the port (i.e. the trunk) is placed back on hook.

**TRUNK HUNTING** Switching incoming calls to the next consecutive number if the first called number is busy.

**TRUNK MONITORING** Feature which allows individual trunk testing to verify supervision and transmission. You dial an access code and then the specific trunk number from the attendant console.

You want the ability to test a specific trunk because normally you might be only accessing a trunk group when you dial an access code. Thus, each time you dial into the trunk group, you might end up on another individual trunk. Some switches have a variation of trunk monitoring, whereby if a user encounters a bad trunk, he can dial a specific code, then hang up. The switch recognizes these digits and makes a trouble report on that specific trunk, possibly reporting it to the operator, keeping it in memory for later analysis or dialing a remote diagnostic center and reporting its agony.

**TRUNK OCCUPANCY** The percentage of time (normally an hour) that trunks are in use. Trunk occupancy may also be expressed as the carried CCS per trunk.

**TRUNK QUEUING** A feature whereby your phone system automatically stacks requests for outgoing circuits and processes those requests on, typically, a first-in/first-out basis.

**TRUNKS IN SERVICE** The number of trunks in a group in use or available to carry calls. Trunk in service equals total trunks minus the trunks broken or made busy for any reason.

**TSAPI** TSAPI stands for "Telephony Services Application Programming Interface." It is a computer-telephone protocol for links between Novell LAN products and telephone switches, and later became Versit's proprietary API. Does for NetWare what TAPI does for Windows.

**TSR** Telephone Service Representative or Telephone Sales Representative.

**TURNKEY** A system that is ready for your use from the moment of purchase (or once the installation is complete). Also, a system that is purchased with all the add-ons and accessories it needs to make it complete and functional from the moment of purchase. The idea is you just turn the key and it works.

**TWISTED PAIR** Wiring that consists of two small, insulated copper wires twisted around each other. When shielded, it reduces interference from other wires and is used in LANs and workstation connections. Unshielded twisted pair is commonly used in telephone cable.

**TWO-PRONG JACK** This type of connection between the handset or headset and the body of the telephone is most commonly found at the receptionist's or attendant's console or an ACD telephone set. Compare to MODULAR JACK.

**TWO WAY TRADE** A shift schedule trade in which two employees are working each other's schedules.

**UCD** Uniform Call Distributor. A device for allocating incoming calls to a bunch of people. Has many fewer features than an Automatic Call Distributor. For a more comprehensive explanation see UNIFORM CALL DISTRIBUTOR. Rarely if ever seen anymore.

**UN** Industry jargon for UNreachable, an unsuccessful call where the agent is unable to speak to the contact or decision maker.

**UNACD** See SERVER-BASED ACD

**UNATTENDED CALL** Calls placed by a computerized dialing system in anticipation of an agent being available to answer the call. A called party is detected answering the phone and no agent is available to serve the call. The system hangs up on the party so as not to create any greater nuisance than has already occurred.

**UNDERSTAFFING LIMIT** The percentage by which you'll allow the scheduling process to fall short of the required staffing level in any period. This typically provides more economical coverage during the least-busy periods of the day.

**UNIFIED MESSAGING** In the past, all messages were different. A voice mail message was recorded and stored on one server. An e-mail came in through a separate date network and sat on a different server. Faxes were physical pieces of paper, or at best, graphics files processed by your fax modem.

Unified messaging brings them all together, if not into one medium, under one roof. It's a class of applications for business that give a user a central location for viewing and dealing with all the incoming messaging traffic he or she might get, both inbound and outbound. One advantage is that it lets the user classify messages by person, so all communication from a particular person can be prioritized up or down. It also brings new tools into the picture, like hybrids that read e-mail aloud, or that OCR fax traffic into e-mail format. This can be very confusing to deal with, but since the average person in American business gets upward of 200 messages a day (no lie), unified messaging, sensibly implemented, can be a smart way to improve productivity. It works best when there's a contact management program behind it, smartly popping up phone numbers and other information when a message comes in. And if it's corporate-wide, then unified messaging can give you an audit trail of incom ing customer messages.

**UNIFORM CALL DISTRIBUTION** See UNIFORM CALL DISTRIBUTOR.

**UNIFORM CALL DISTRIBUTOR** A device for distributing many incoming calls uniformly among a group of agents. A Uniform Call Distributor is generally less "intelligent," and therefore less costly than an ACD. A UCD will distribute calls following a predetermined logic, for example "top down" or "round robin." It will not typically pay any heed to real-time traffic load, or which agent has been busiest or idle the longest. Nor will it allow you to use skill-set routing, or any sophisticated overflow patterns.

Also, a UCD's management reports tend to be rudimentary, consisting of simple peg counts, as opposed to an ACD, which can produce reports on the productivity of agents.

**UNIVERSAL AGENT** A telephone agent who answers incoming calls and also makes outgoing calls. Also sometimes referred to as a blended agent.

For a long time, agents were either just just "inbound" or just "outbound" — because managers felt that most agents were not capable of doing both. And because it was difficult to manage people who were floating from one side of the fence to the other. The skills were, allegedly, too different. Now the idea is to "empower" the agent with more flexibility and make them "universal," i.e. capable of being used for both inbound and outbound. What's really happening is that since you can often move an agent from inbound to outbound at the touch of a button, there are cost-savings to be had from cross-training and from the added staffing flexibility that gives you.

The term is registered to Melita International, and in that case refers specifically to agents with inbound and outbound capability, plus access to other resources. The term is also used generically, and widely.

**UNIVERSAL DEVICE** An SCSA device. A call processing device which has every conceivable resource for the handling of calls. The SCSA programming applies resources from many different physical devices to a call processing task. These then act as if they were a single universal device.

**UNIVERSAL PORTS** A modern telephone system is typically an empty cabinet into which you slide printed circuit cards.

In the old days, phone systems had dedicated slots — meaning you could only slide one type of printed circuit card into that particular slot. As phone systems got more advanced, they acquired "universal ports." Our definition of a universal port is that all the slots are totally flexible — namely that you can slide any trunk or phone card (either electronic or single line phone) into any slot in the phone system. The advantage of this is obviously a far more flexible phone system, able to accommodate lots of phones and few trunks or vice versa.

**UNLISTED NUMBER** A telephone number that is not printed in a telephone directory, at the request of the person (or people) using that number. Making a "cold call" to a consumer with an unlisted number, no matter how you got that number, is tricky business. Over 25% of private phone numbers in major metropolitan areas are now unlisted — a "service" their subscribers pay extra for. See NON-PUBLISHED.

**UNPBX** A telecommunications server or a "PBX in a PC." Actually, to be an UnPBX, the platform doesn't have to be a PC, it can be any data processing platform, but PC-based PBXs seem to be the most common form of UnPBX. Because UnPBXs are built on open platforms, with open standards, it is very easy to add third-party ACDs, IVR systems, voice mail, computer telephony systems, LAN and Internet connections and other multimedia applications to the system.

The drawback to these systems is that they are only are reliable as the platform they are built on, and telecommunications applications are much less tolerant of delays than your average computer application. How many times have you entered something into a word processing program or a spreadsheet, only to have your computer's hard drive purr along, ignoring your input as if it was thinking of something else? How many times has your computer crashed? Imagine these things ending all the telephone conversations going on in your company at the time. It's a scary thought.

The solution, of course, is more reliable platforms, and the industry is working on them.

This is how Dialogic defines an UnPBX: an UnPBX generally refers to a business telephone system or PBX that is built using open components like a PC, a standard operating system, and CT hardware like trunk, station interfaces, and voice processing boards. The vendor writes applications for the same telephony capabilities found in PBXs and key systems, plus other core business applications like voice mail.

While an UnPBX offers many advantages over a traditional PBX because of the use of open components (Dialogic goes on), available products today are typically implemented using a closed approach — limiting users to a single supplier. System owners have little choice of applications and slower access to new technologies, simply because the software comes from one supplier and is written to a specific selection of hardware components. Although some suppliers do offer an open interface like TAPI, the buyer is still limited to the suppliers' specific implementation and the technologies that the system vendor can support.

For call centers, the UnPBX is important because it is the basis of the UnPBX ACD, which is an UnPBX with an ACD application added on — an important technology for small call centers with sophisticated technical needs — and the server-based ACD, which could be called an UnACD.

**UNPBX ACD** See SERVER-BASED ACD.

**UNSUCCESSFUL CALL** A call attempt that does not result in the establishment of a connection.

**USAGE-BASED** "Usage-based" refers to a rate or price for telephone service the depends on volume consumed, rather than a flat, fixed monthly fee. Until a few years ago, most local phone service in the United States was charged on a flat rate basis. Increasingly, phone companies are switching their local charging over to usage-based. Flat-rate calling will probably disappear within a few years. Allegedly, usage-based phone service pricing is fairer on those phone subscribers who don't use their phone much. Usage-based pricing is not consistent throughout the U.S. Typically, you get charged for each call. And the charging is very much like that for long distance — by length of call, by time of day and by distance called.

**V-COMMERCE** V-commerce was concocted by Nuance, a speech recognition company, and promptly trademarked. Normally trademarking a term is a foolhardy practice if you want it to become a common coinage. Nonetheless, v-commerce refers both to a group of associated vendors Nuance has gathered to develop applications for their speech recognition system, and to the idea of speech-enabled-e-commerce.

V-commerce is a way of shorthanding the notion of telephone transactions enhanced by a speech recognition front end. It's somewhat evocative. It's our belief that the practice of adding speech rec to call center apps will become so prevalent in the next few years that we won't need a separate term to describe it; it will just be a fact of the way call center call flow will be designed.

As an example, note that we don't have (or need) a separate term to describe a call center frontended by an IVR or traditional voice response system.

**VAD** Value Added Dealer. Another term for Value Added Reseller (VAR). Essentially, VARs or VADs are companies who buy equipment from computer or telephone manufacturers, add some of their own software and possibly some peripheral hardware to it, then resell the whole computer or telephone system to end users, typically corporations.

**VANITY NUMBER** A phone number that spells a word, like the name of the company or product. They are popular with 900 and toll-free services, like 800-MAT-TRES for Dial-A-Mattress, 900-ROBOCOP to promote a movie, or 800-USA-RAIL for Amtrak fares and schedules. Names or cute phrases are easy for customers to remember, but hard to dial. When seeking a new vanity number, remember to ask for it in all the new toll-free codes; you don't want your competitors grabbing your cool phone number in 888 or 877 because you forgot.

**VAR** Value Added Reseller. A company that buys a piece of equipment from a manufacturer, adds their own software, peripheral or other enhancements to it, and then resells it to the end user. The enhancements often streamline the product for a particular industry niche. VARs include business partners ranging in size from providers of specialty turn-key solutions to larger system integrators.

**VARIABLE CALL FORWARDING** A feature of some toll-free services. It lets you route calls to certain locations based on time of day or day of week.

215

**VDN** Virtual Defined Network. A Lucent term for the network that can be defined between or among Definity switches.

**VECTOR** "A series of commands or call processing steps that determine how calls are handled or routed. Call vectoring offers more flexibility in managing incoming call traffic, assuring that calls get answered quickly, by the best qualified agent," according to AT&T.

**VERY LOW FREQUENCY MAGNETIC RADIATION** See VLF.

**VIDEO KIOSK** It isn't practical to expect everyone who wants to communicate with your call center visually — that is, by video — to have all the equipment they need at home. A video kiosk provides that equipment in a public area, such as a mall or building lobby, or in a semi-public area such as in a bank branch or even a supermarket.

Though this technology hasn't caught on in a big way, it's being explored extensively by financial services firms, who are using the ATM-use model to perhaps justify consumers going up to a video kiosk and securing a loan, or an insurance policy — something you need human interaction to accomplish, but that can be done more cheaply when the humans are centrally located.

**VIRTUAL BANDING A WATS** service that provides tiered pricing, but not physical bands that require separate trunks. With virtual banding any call can go out on any trunk, and you are charged based on the destination of the completed call. See BAND, WATS.

**VIRTUAL BYPASS** Virtual bypass is a way smaller users can fill the unused portion of local T-1 dedicated loops going from a user site to a local office of a long distance company, called a POP (Point of Presence).

**VIRTUAL CALL CENTER** A "virtual call center" is several groups of agents, usually in geographically separate locations, that are treated as a single center for management, scheduling and call-handling purposes. In some rare cases the virtual call center is made up of agents working from their homes, with a telephone switch at company headquarters (or Centrex) routing the calls.

All virtual call centers require transmission services between sites that is more than the average local telephone company offers. To switch the calls, call routing information, and data between the sites requires a lot of bandwidth. This bandwidth is achieved with a private line between locations, switched high bandwidth telephone services or ISDN. A virtual call center also requires sophisticated routing and networking features from all the call centers' ACDs.

**VIRTUAL NETWORK** A network that is programmed, not hard-wired, to meet a customer's specifications. Created on as-needed basis. Also called Software Defined Network by AT&T.

**VIRTUAL LAN** A logical grouping of computer users regardless of their physical locations on the network. It can also mean a LAN that has been extended beyond its geographical limit through T-1 or Sonet services.

**VIRTUAL PRIVATE NETWORK VPN.** A service provided by a local or long distance telephone carrier that offers the special features of a private network (conditioning, error testing, and higher speed, full-duplex, dynamically allocated bandwidth), over the public switched network, usually for a cost that is lower than installing a real private network.

**VLF** Very Low Frequency magnetic radiation. A slice of the electromagnetic spectrum emitted by video display terminals (aka computer screens) whose long-term danger to human beings has been neither proven or disproven. Emissions are greatest to the rear and sides of the display. Something to keep in mind when designing the layout of your call center.

**VOCABULARY DEVELOPMENT** Development of specific word sets to be used for speaker independent recognition applications.

**VOICE ANNOUNCEMENT SYSTEM (VAS)** Card that provides delay announcements to callers waiting for available agents, and call- type announcements to agents.

**VOICE BOARD** Also called a voice card or speech card. A personal computer printed circuit board (expansion card) that performs voice processing functions. A voice board has several important characteristics: It has a computer bus connection. It has a telephone line interface. It typically has a voice bus connection. At a minimum, a voice board includes support for going on and off-hook (answering, initiating and terminating a call); notification of call termination (hang-up detection); sending flash hook; and dialing digits (touchtone and rotary). See VOICE BUS and VRU.

**VOICE BUS** A circuit that controls the voice processing cards within a computer system. Physically, the bus is often a connection (tiny pins) at the top of a voice processing card. A ribbon cable connects each card. There are several voice bus "standards." Two come from Dialogic. One is called AEB, Analog Expansion Bus. And one is called PEB, PC Expansion Bus (a digital version). One comes from a consortium of companies and is called MVIP. A voice bus gives you flexibility to mix and match voice processing boards, like voice recognition, voice synthesis, switching, voice storage into a single system.

**VOICE CARD** See VOICE BOARD.

**VOICE MAIL** Voice mail is a device which lets your receive, edit, forward, listen to, create and change recorded voice messages, almost always in conjunction with a telephone system of some kind. Voice mail is an extremely important telecommunications concept for most businesses. For call centers, there are just a few things of note.

First, an inbound call center's voice mail needs are completely different from the needs of any other business. Voice mail was created for communication between individuals. The very nature of a call center implies that any one of several agents is capable of handling a given call, and efficiency takes precedence over individuality. An inbound call center requires a voice mail system that can deliver the recorded voice message to the next available agent when calling volume per-

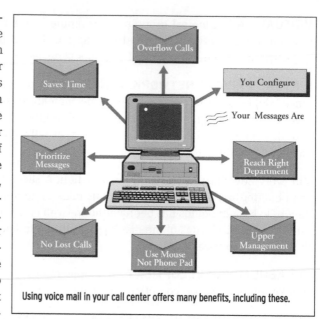

Using voice mail in your call center offers many benefits, including these.

mits. There are ACD systems that provide this feature.

In some inbound call centers, agents have their own accounts and clients and the individual-to-individual messaging concept is valid. These call centers can use standard voice mail systems. These centers and agents must be aware of voice mail as a customer service tool. Many people hate voice mail, and for good reason. Their personal or business needs have been hurt by voice mail abuse: unnecessarily screening calls, outdated or non-informative greeting messages, not returning calls promptly, not returning calls at all. Your call center should have checks and balances in place to make sure your agents are using voice mail to serve your customers, not as a tool to ignore them.

For outbound call centers, dealing with voice mail on the other end of the call is the important issue. Some sales experts feel that leaving a voice mail message on a cold call is a waste of time. Chances are slim that your prospect will return your call. Other experts advocate leaving a message that will entice the prospect to return your call for more information. The idea is to leave a brief (30 seconds or less) advertisement for your product, service, or the call back. "I believe I can save your company 20% on office supplies. Please call me back so we can determine if our program will work for you." Agents should always include their name, your company name and a telephone number.

Voice mail systems can be stand-alone, PC-based or integral to a telephone switch. Their are voice messaging service bureaus that rent out parts of their system by the mailbox. We would even consider an answering machine a voice mail system. Prices range from $30 for a real cheap answering machine to

$200,000 or more for an extensive, sophisticated system that can handle messages for a large corporation.

Here are some of the benefits of voice mail:

1. No more "telephone tag." Voice mail improves communications. It lets people communicate in non-real time.

2. Shorter calls. When you leave messages on voice mail, your calls are invariably shorter. You get right to the point. Live communications encourage "chit chat" — wasting time and money.

3. No more time zone/business hour dilemma. No more waiting till noon (or rising at 6 A.M.) to call bi-coastally or across continents.

4. Fewer callbacks. In some cases, as many as 50%.

5. Improved message content. Voice mail is much more accurate and private than pink slips. Messages are in your own voice, with all the original intonations and inflections.

6. Less paging and shorter holding times.

7. Less peakload traffic.

8. 24-hour availability.

9. Better customer service (if used properly).

10. Voice mail allows work groups to stay in contact — morning, noon and night.

**VOICE MAIL JAIL** The dead-end you find yourself in a voice mail system, when no matter what you try, you can not reach a live human being. Commonly this is caused when pressing "0" on your telephone keypad does not deliver you to an operator or other human. Sometimes it is caused when one message says, "If you need to speak to someone right away, call Jane at 334." When you dial that extension, Jane tells you to talk to someone else, and so on, until you are just stuck. Don't put your customers in voice mail jail. Always give them a way to reach a human. (And coordinate contingency messages so you don't create a "chain gang.")

**VOICE MAIL SYSTEM** A device to record, store and retrieve voice messages. There are two types of voice mail devices — those we will call "stand-alone" and those that come with a telephone switch like an ACD or a PBX. Often these switch-based voice mail systems are simply the guts of a stand- alone system mounted in the phone system cabinet. Once, "stand- alone" system did not integrate with your phone system. Today a voice mail system that cannot activate a message-waiting indicator or automatically transfer a call from the phone system to the voice mail system after a certain number of rings is obsolete.

There are several levels of switch/voice mail integration though. Some voice mail systems can alter the display on your telephone sets "soft" feature buttons to make accessing voice mail playback features such as forward, reverse, slow, fast, stop as simple as pressing the appropriate button. See VOICE MAIL.

**VOICE MESSAGING** Recording, storing, playing and distributing phone messages. Essentially voice messaging takes the benefits of voice mail (such as bulk messaging) beyond the immediate office to almost any phone destination you select. Voice messaging is often done through service bureaus. Here's one vendor's interesting way of looking at voice messaging: break it into four distinct areas: 1. Voice Mail, where messages can be retrieved and played back at any time from a user's "voice mailbox"; 2. Call Answering, which routes calls made to a busy/no answer extension into a voice mailbox; 3. Call Processing, which lets callers route themselves among voice mailboxes via their touchtone phones; and 4. Information Mailbox, which stores general recorded information for callers to hear.

**VOICE-OVER-IP** Commonly referred to as Internet Telephony. What if there were a way to make a phone call that could bypass some or all of the telephone network? There is, and it involves using data networks and the internet transmission protocol. A phone call, consisting of voices, can be broken apart and sent across a data network as a series of packets, just like any other data transmission. With the right hardware and software, it can be reassembled on the other end with enough accuracy for two people to have a "phone call" in real time.

The first implementations of this involved PC-to-PC calls — someone in New York would talk to someone in London by loading the software on his PC, initiating a call request which would alert the person on line in London (with his software also running). This is cumbersome and technologically inferior to making a POTS call between those two points, but it does have one very important advantage: it's effectively free, because the entire call travels over the TCP/IP network.

The next logical step was to use a combination. Instead of sending a call over one network or the other, it can use a combination. Start the call on the telephone, using the local loop; then at the switch, it can be sent over the longest leg by packet network, over the Internet, whatever is cheapest and technologically feasible at that point. It can be reassembled at the remote switch, transferred to the local carrier at their end point, and routed to a traditional telephone number, all at a fraction of the cost of traditional voice telephony.

The one disadvantage to internet telephony so far has been quality; because the call is broken into packets for transmission, it can sometimes have a bit of latency. (This is the echo-like sound that used to occur with calls bounced off a satellite — annoying, but not a conversation killer.) This has been reduced quite a bit lately, but still seems to happen, particularly with extremely long distance calls and at times when network traffic is high.

It is thought that the long distance carriers will eventually shift a lot of their network traffic to the internet or some post-internet packetized network technology. Indeed, there is no reason why any fax traffic has to travel over the old system - latency isn't an issue, but cost is. See FAX-OVER-IP.

**VOICE-OVER-NET** See INTERNET TELEPHONY, VOICE-OVER-IP.

**VOICE PROCESSING** An umbrella term that covers all the technologies that allow you to record, retrieve and manipulate the spoken word, especially over telephone lines or in conjunction with a telephone switch. The technologies include: audiotext (voice bulletin boards), automated attendant, interactive voice response, text-to-speech, voice mail, voice recognition, voice response and those gizmos that play a three- second greeting replacing an agent's live greeting. For more information about these individual technologies, see the entry under the name of that technology.

Voice processing is a broad function made up of two narrower functions: call processing and content processing. Call processing consists of physically moving the call around. Think of call processing as switching. Content consists of actually doing something to the call's content, like digitizing it and storing it on a hard disk, or editing it, or recognizing it (voice recognition) or using it as input into a computer program.

**VOICE RECOGNITION** Another way of saying "speech recognition," which is the preferred term. Both are the ability of a machine to recognize words spoken by a human. For most people, the two terms are identical. People in the speech recognition industry use only "speech recognition," however. They are tired of "voice recognition" being confused with various voice processing technologies, especially "voice response." There are two kinds of speech recognition. One works with only one speaker who has trained the system (speaker dependent), the other works with any person who uses the system (speaker independent). See SPEECH RECOGNITION, SPEAKER DEPENDENT, and SPEAKER INDEPENDENT.

**VOICE RESPONSE UNIT** VRU. A term that can refer either to an interactive voice response unit, an automated attendant or a very simple, "play a message and pass the call on" unit. The term is commonly used in call centers to mean the automated voice system that greets the caller before the caller gets to a live agent. In some cases this is an IVR system that deals with the entire transaction. In other cases it is an IVR system that prompts the caller for an identification number, pulls up a customer record, then routes the call based on that information and presents both call and information to the agent.

Or, the caller may be greeted by an automated attendant, which offers the caller a choice of departments or product lines and routes the call accordingly. Mostly simply, it can be a recorded message that the caller hears before being routed to an agent or put in queue. See IVR and FRONT-END.

**VOICE STORE AND FORWARD** Another, older, name for voice mail. See VOICE MAIL.

**VOICE TALENT** The pleasant-sounding person who records a message-on-hold, IVR prompt, automated attendant greeting or other important, frequently listened to voice processing message. The term implies a paid professional, but this person can be your receptionist or an agent with a particularly nice voice. The voice talent is the voice of your company. Choose well and make sure the person will be available in the future to record minor changes in your prompts or messages. Believe it or not, hiring a professional is a good way to do this.

**VRU** See Voice Response Unit. Also see FRONT-END and IVR.

**VOICE SERVER** A special PC sitting on a LAN (Local Area Network) that contains voice files which are accessible by the other PCs on the LAN. The server can be a part of a voice mail system, a fax-on-demand system, a voice-based e-mail system or perform various other voice processing functions. The idea is it is a special PC designed to make the tricky task of storing and retrieving voice information easier. Keep in mind that Digital Sound Corporation has a federal trademark on the term "Voiceserver," as one word. This brings up a whole can of worms on the trademarking process, which has a giant loophole which lets companies trademark technology terms not used by the general public. But we won't bore you with that right now.

**VSE** A British Term. Voice Services Equipment, a generic term for voice response unit, interactive voice response, voice processing unit and so on.

**VPN** See VIRTUAL PRIVATE NETWORK.

**VRU** See VOICE BOARD and VOICE RESPONSE UNIT.

**VRU CONNECT SIGNALING** When the incoming call is answered by a VRU, a caller typically inputs information through a touchtone keypad. This user-defined information is now "whispered" as an audio announcement by the VRU to a service representative. After the VRU has conveyed the caller's information to the agent, the VRU drops out and the call is transferred to the agent. The most notable benefits are 1) enabling agents to better anticipate and consequently address the needs of callers, 2) eliminating the need for agents to request information already gathered by the VRU, 3) presenting a better organized and professional appearance to callers, and 4) expediting the transaction process.

**WAN** See WIDE AREA NETWORK.

**WATS** Wide Area Telecommunications Service. Basically, a discounted toll service provided by all long distance and local phone companies. It's a generic term. There are two kinds of WATS service: *in* and *out*. In is your typical inbound 800 toll free service.

Out is now commonly regarded as long distance and in fact some long distance companies still call their long distance service WATS. But WATS used to refer to the practice of breaking territory into "bands" of states for billing purposes. You'd buy service by the band.

**WEIGHTED AVERAGE** A method of averaging several numbers in which some numbers are increased before averaging because they have more significance relative to the other numbers.

**WEIGHTED CALL VALUE** WCV. The average handling time of a call transaction. ACD vendors count this differently. Typically, it's a combination of the talk time and the after-call work or wrap-up time.

**WET T-1** A T-1 line with a telephone company-powered interface.

**WHISPER PROMPT** The verbal announcement, silent to the caller, that tells the agent what type of call is arriving. Less sophisticated than a "screen pop" that delivers a computer screen of information to the agent, a whisper queue, or whisper prompt, can be accomplished by the telephone system (usually an ACD) alone. A screen pop requires integration between the telephone system and the computer system.

**WHISPER QUEUE** See WHISPER PROMPT.

**WHISPER TECHNOLOGY** A call comes into a call center. The voice response unit prompts the caller to the enter their account number. When the call is transferred to the agent, the VRU "whispers" the account number to the agent, who then manually types it into his computer. This technology is now obsolete, since VRUs can now transfer their account number directly into the agent's database and have the look up done automatically. And the call is transferred simultaneously.

**WHITE NOISE** A signal whose energy is uniformly distributed among all frequencies within a band of interest. Seldom occurring in nature, white noise is a useful tool for theoretical research. White noise is also used less scientifically to simply mean background noise. When the first digital PBXs came out, their intercom circuits were so "clean," they spooked users who were used to some noise on the line. And some PBX manufacturers added a little "white noise" to their PBXs.

**WIDE AREA NETWORK** WAN. A data network typically extending a LAN (local area network) outside the building, over telephone common carrier lines to link to other LANs in remote buildings in possibly remote cities. A WAN typically uses common-carrier lines. A LAN doesn't. WANs typically run over leased phone lines — from one analog phone line to T1 (1.544 Mbps). The jump between a local area network and a WAN is made through a device called a bridge or a router. Bridges operate independently of the protocol employed.

**WIDE AREA TELEPHONE SERVICE** See WATS and 800 SERVICE.

**WIRE CENTER** The location where the telephone company terminates their local lines with the necessary testing facilities to maintain them. Usually the same location as a class 5 central office. A wire center might have one or several class 5 central offices, also called public exchanges or simply switches. A customer could get telephone service from one, several or all of these switches without paying extra. They would all be his local switch.

**WIRE CENTER SERVING AREA** That area of an exchange served by a single wire center.

**WIRED-FOR CAPACITY** The wired-for capacity represents the upper limit of capacity for a particular configuration. To bring to a phone system to its "wired-for capacity," all that's necessary is to fill the empty slots in the system's metal shelving (its cage) with the appropriate printed circuit boards. "Wired-for Capacity" is a marginally useful term, giving little indication of the type of printed circuit boards — trunk, line, special electronic line, special circuit, etc. — that can be installed.

**WIRELESS HEADSET** A headset that uses a similar technology to a cordless home telephone. Manufacturers use "wireless" instead of "cordless" to avoid the stigma of the cheapy cordless phones that don't work very well. There are two types of wireless headset: one with a portable battery pack put in a pocket or on a belt, with a wire connecting the headset itself, and one where the battery pack is in the earpiece. The one-piece headset is heavier, but is probably more convenient. The two-piece model lets you use your existing headset, and is significantly lighter on your head.

Because they work on radio frequencies, interference prevents wireless headsets from being used in large call centers. In very small call centers, in informal call centers where each agent has his or her own office (inside sales) or in a small technical support center, a wireless headset lets agents walk around to access information, demonstration equipment or just stretch their legs.

**WORD SPOTTING** In speech recognition over the phone, word spotting means looking for a particular phrase or word in spoken text and ignoring everything else. For example, if the word to spot was "brown," then it wouldn't matter if you said "I want the brown one," or "how about something in brown?" In short, word spotting is the process whereby specific words are recognized under specific speaking conditions (i.e. natural, unconstrained speech). It can also refer to the ability to ignore extraneous sounds during continuous word recognition.

**WORK AT HOME** The new term for telecommuting. Between the Internet and software-based ACDs, the technology for telecommuting has improved greatly over the last 10 years. Vendors of this next generation technology don't want you thinking about that failed telecommuting project of yore when you consider their product, so they call it a work-at-home system. Call center agents are particularly well suited for working at home (compared to say, a waiter or a football player). When one of your agents works at home, he or she is called a REMOTE AGENT.

**WORKFLOW** The way work moves around an organization. It follows a path. That path is called workflow. Here's a more technical way of defining workflow: The automation of standard procedures by imposing a set of sequential rules on the procedure. Each task, when finished, automatically initiates the next logical step in the process until the entire procedure is completed.

**WORKFLOW MANAGEMENT** The electronic management of work processes such as forms processing (for example, insurance policy acceptances) or project management using a computer network and electronic messaging as the foundation.

For example, let's say when an order comes into your company, it first goes to the credit department, then to the warehouse where inventory is checked, a letter sent if something is not in stock, products are picked and packed, then sent to the shipping department. Each department might be responsible for bringing the order to the next step, or the process might be done automatically. Either way, the process is called workflow management.

Workflow management can be automated and when all the work is done on computers or electronically in some way, having an automated workflow management system can be a big benefit. In your call center automated workflow management means that when an agent completes an order it automatically is sent to the next step, whether that's the credit department, or somewhere else. With automated workflow management a field salesperson can hit a button to assign a phone salesperson to schedule an appointment with a client. Or the warehouse can assign an agent to call a customer and tell her that certain items in her order are not available. The system automatically sends back the confirmation that the work has been done and other details (such as the day and time of the appointment to the salesperson).

**WORKFORCE MANAGEMENT** Call center workforce management is the art and science of having the right number of people...agents...at the right times, in their seats, to answer an accurately forecasted volume of incoming calls at the service level you desire. Naturally there is a whole class of software that accomplishes this task, much of it quite good.

**WORKLOAD** The total duration of all calls in a given period (half hour or quarter hour), not counting any time spent in queue. This figure is equal to the number of calls times the average handle time per call.

**WORKSTATION** In the telecom industry, a workstation is a computer and a telephone on a desk and both attached to a telecom outlet on the wall. The computer industry tends to refer to workstations as high-speed personal computers which are used for high-powered processing tasks like CAD/CAM or engineering. A common PC is not usually considered a workstation. (Unless it's running Unix, or is powered by a RISC chip. Then, for some reason, it seems to qualify.) The term workstation is vague.

**WORLD NUMBERING ZONE** One of eight geographic areas used to assign a unique telephone address to each telephone subscriber.

**WRAP-UP** Between-call work state that an ACD agent enters after releasing a caller. It's the time necessary to complete the transaction that just occurred on the phone. In wrap-up, the agent's ACD phone is removed from the hunting sequence. After wrap-up is completed, it is returned to the hunting sequence and is ready to take the next call.

**WRAP-UP DATA** Ad hoc data gathered by an agent and entered into either the ACD system or some other customer data management application following a call.

**WRAP-UP TIME** The time an employee spends completing a transaction after the call has been disconnected. Sometimes it's a few seconds. Sometimes it can be minutes. It depends on the rules for the bcenter. XDP Ron Stadler of Panasonic dreamed this one up. It stands for eXtra Device Port. It's an analog RJ-11 equipped port on the back of a Panasonic digital telephone, which is driven by Panasonic's Digital Super Hybrid switch. The XDP is an extension line completely separate from your digital voice line. You can be speaking on the phone while receiving or sending a fax or while sending or receiving data. Or plug a cordless phone or answering machine into the XDP.

**ZERO USAGE CUSTOMER** A carrier term for a customer who has not placed a call over the network, even though he/she is an active customer. Sometimes used interchangeably, but incorrectly, with the term "no usage customer."

**ZIP TONE** Short burst of dial tone to an ACD agent headset indicating a call is being connected to the agent console.